Practical Cytology

Practical Cytology

R. G. W. TAYLOR
St Luke's Hospital, Bradford, Yorks

1967

ACADEMIC PRESS
London and New York

ACADEMIC PRESS INC. (LONDON) LTD.
BERKELEY SQUARE HOUSE
BERKELEY SQUARE
LONDON W.1.

U.S. Edition published by
ACADEMIC PRESS INC.
111 FIFTH AVENUE
NEW YORK, NEW YORK 10003

Copyright © 1967 by ACADEMIC PRESS INC. (LONDON) LTD.

All Rights Reserved
NO PART OF THIS BOOK MAY BE REPRODUCED IN ANY FORM BY
PHOTOSTAT, MICROFILM, OR ANY OTHER MEANS, WITHOUT WRITTEN
PERMISSION FROM THE PUBLISHERS

Library of Congress Catalog Card Number: 67–24317

Printed in Great Britain by
T. & A. Constable Limited, Edinburgh

"Read not to contradict and confute, nor to believe and take for granted, nor to find talk and discourse, but to weigh and consider."

Of Studies
Francis Bacon 1561–1626.

Acknowledgments

Firstly I wish to acknowledge the tutelage of Dr R. A. McInroy, Consultant Pathologist and Head of this Cytology Unit, whose tolerance and counselling have allowed me an unbiased initiation to this study.

The photographs were prepared with the assistance of Mr Peter Harrison, Hospital Photographer, Bradford "A" Group Hospital Management Committee and Mr David McIlreavy, Gillet and Sibert Ltd.

I wish to thank my colleagues, Dr Kathleen Horne and Mr William Kernohan for their help and encouragement, and also Mrs E. Weston and Miss Maureen Millward who have patiently typed the script.

Dr McInroy has kindly read through the manuscript, but, of course, any errors and omissions are my own.

Finally I would like to thank my publishers for presenting the book.

Preface

This book is offered primarily as a guide to medical laboratory workers contemplating the field of cytology. Sufficient basic information has been included to allow its use as a bench manual, but the book is intended to give incentive to consult standard works on histology and cytology. Two suggested volumes are "A Textbook of Histology", by Maximow and Bloom, and "The Cytologic Diagnosis of Cancer", by Graham. It behoves every worker entering the field to ensure that he first has an insight into the histology of the relevant tissues and subsequently to correlate his cytological findings with resulting histology. Only by this means will he be able to discover the source of the cells present in the smears he examines and so produce a constructive interpretation on what he sees. A cytology service based on screening by "rule of thumb" criteria is bound to yield inaccurate results with a disproportionate number of false positives, false negatives or a mixture of both.

The information given on the normal and abnormal cell does not follow the conventional presentation of medical cytology, but the observations offered in this regard are based on the author's accumulated experience acquired during the examination of more than 50,000 slides and during the training of others. Most of the abnormal cells seen will be due to causes other than neoplasia, and an ability to differentiate between these atypical cells requires great care and considerable experience. Absolute accuracy in the identification of every cell seen is an ideal that will never be attained with present staining techniques.

Initiates will find much to interest them; solutions to many of the problems that present themselves have still to be found. Examples are the effect on epithelia of fluctuations in the serum level of various hormones and the changes in form and function that can occur in apparently committed cells.

Wide differences of opinion exist between many workers in pathology concerning the staffing of cytology departments. To those who may be consulted on this problem there are two pieces of advice that may prove helpful. Firstly, to remember the adage "two heads are better than one"; the consensus of several experienced workers' opinions will usually yield the correct interpretation of difficult smears. Secondly, to realize that this work is best carried out in an atmosphere of absolute calm; the constant hurly-burly of the average clinical pathology department is not conducive to accurate cytology.

R. G. W. Taylor

January 1967

Introduction

Several early workers have been credited with the original practice of applying cytological methods as an aid to diagnosis. Probably the first observations were those made during the early renaissance in medicine. Microscopic examination of urine was reported, although the interpretation placed on the cells seen is obscure. Today we regard this examination as of extreme importance.

In pathology, "exfoliation" is a word used to describe the loss of cells occurring at the surface of tissues exposed to desiccation or abrasion. It is a natural process and the cells lost were originally an integral part of the tissue. Many of the cells we see in cytological smears are not exfoliated. They reach the surfaces from which they may be harvested by other methods, principally by virtue of their own capability to move through tissues in response to various stimuli. The loss of plaques of *squamous* epithelial cells from a surface is described as *desquamation*, and the term embraces both the normal and abnormal losses of these cells. It should not be used to describe the loss of other types of epithelial cells.

A resurgence of interest in cytology was initiated by an incidental observation made by Papanicolaou. During a study of the effect of oestrogens on vaginal epithelial cells he noticed that the abnormal cells shed from a neoplasm in the cervical epithelium are sufficiently characteristic to indicate the lesion. This is particularly true when they originate in the squamous epithelium. More important, the cells may be seen when the tumour is not obvious clinically, by using such routine aids as are otherwise available. However, the proportion of atypical cells seen in the smear bears little relationship to the size of the lesion; neither is it possible cytologically to determine whether any of the cells have invaded the underlying tissue. It can be said, however, that smears produce a typical "positive" picture in most cases of pre-invasive carcinoma when surgical intervention has a good chance of effecting a cure.

Since, at this stage, the lesion is not visible, it is largely chance whether a simple biopsy will contain the site of origin of the cells. Even if it does, a random section taken from the block may not demonstrate the pathology. These possibilities can be eliminated by taking the whole of the epithelial surface that is likely to contain the lesion, e.g., the "cone biopsy of cervix", and processing the total tissue in an orientated series of blocks.

The bias toward gynaecological cytology in this volume reflects the lines along which cytology has expanded in this and other hospitals. Often the initial demand is for a cytology service designed to detect otherwise occult squamous lesions of the uterine cervix and so allow early histological investigation. However, when the service is applied to a population survey of "well women" at risk, the possible presence of adenocarcinoma must be

INTRODUCTION

remembered. Screening these smears required a particularly conscientious attempt to exclude this possibility. Many of the individuals submitting themselves for investigation regard "cancer of the womb" as an entity. A negative report is equally as important as a positive one. It means that slides at all levels of screening must be examined by a trained worker, who attempts to identify every cell he sees. Instructing a junior to "go through these and screen out the negatives" may produce a reward of false negatives.

Use is also made of cytological screening methods in the detection of urinary bladder carcinoma. This is an industrial hazard to certain workers who are at risk as a result of exposure to carcinogens known to cause this disease. Cytological examination of the urine from such workers is regularly carried out as a means of early detection.

Currently, the weight of informed opinion is against the feasibility of routine periodic screening of individuals at risk to gastric and respiratory neoplasia, holding that no corollary exists to the long (5 to 10 years) *in situ* period occurring in the case of cervical squamous carcinoma.

A consequence of this activity in cytology, as a technique for the early detection of carcinoma, has been the bringing of a fresh and more comprehensive perspective to the cytological examination of other commonly collected body fluids and secretions.

Contents

Preface	*vii*
Introduction	*ix*
Chapter I. Preparation of Smears	1
Chapter II. Staining, Mounting and Examining Smears	7
Chapter III. Structure of the Cell	25
Chapter IV. Gynaecological Cytology	68
Chapter V. Sputa	86
Chapter VI. Serous and Other Fluids	93
Chapter VII. Documentation	105
Chapter VIII. Ancillary Techniques	112
Appendix: Formulae, Notes on Centrifuges and Population Screening for Cancer of the Cervix	133
Bibliography	141
Subject Index	143

CHAPTER I

Preparation of Smears

Standard 3 ×1-in slides are used. For departments outside the laboratory, it is convenient to work with slides having frosted ends. Information may be written directly on these slides with a lead pencil. A hard lead (Grade 5H) is most suitable, since the softer leads tend to smudge if the slides are mishandled. Marking with any of the ball-point type of pencil is washed away during fixation or the staining sequences. Generally speaking, slides should not be coated with albumen or similar adhesives, since these result in an undesirable background stain. Whenever smears are made, immediate fixation is essential. Smears that have been allowed to dry before fixation show characteristic "drying effects"; an attempt to formulate reports on such slides is a laborious process and of uncertain accuracy.

Some of the fluids and secretions collected for cytological examination are of sufficient volume to allow transportation to the laboratory without deterioration of the cells. Concentration procedures may then be carried out before the smears are made. When the volume of aspirate is small, desiccation is a constant risk, and it is these specimens that have to be smeared and fixed at the time and place of collection. This is the practice in nearly all gynaecological cytology work. It is prudent, of course, to request more than one slide in these circumstances, not only to reduce the risk of having a single unsatisfactory specimen to examine but also to allow additional staining sequences to be used when they may be helpful. Routine procedure with hospital gynaecological specimens in this laboratory, for example, includes the preparation of a "Gram" from a vaginal pool smear. The flora present may yield useful information when the Papanicolaou-stained smears are interpreted, and it is a more accurate stain to use for identifying a Monilia infection.

Smears on Cellophane

Investigation in this laboratory has shown that problems in postage and storage of slides, mentioned later, can be overcome in large measure by using Cellophane as the supporting medium, although the method is not in use. A suitable product is P.T.400, obtainable in 1-in wide rolls from British Cellophane Ltd. Smears are made in the usual way on 6-in strips of Cellophane and fixed while still moist. The strips may be numbered with a hard lead pencil: the indentation is easily read and not eradicable.

Fixation

The principal requirement of a cytological fixative is that it should precipitate the cell protein, and in particular the nuclear protein in a pattern characteristic of the form during life. Cytological fixatives used in histological techniques are generally successful although they expose the smear to the risk of loss of material if water is the basic solvent. These are mainly mercury-containing fixatives, an example of which is Schaudin's fluid.

It is more convenient, however, to use a fixative that has a dehydrating

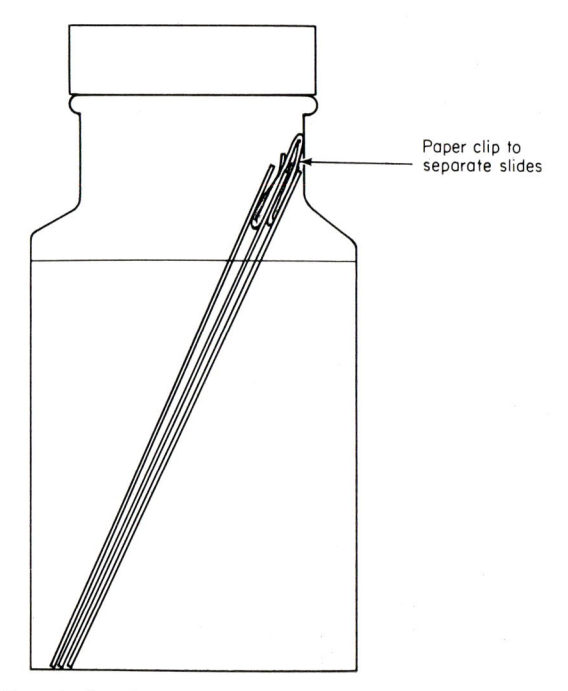

Fig. 1. Specimen container for routine gynaecological smears.

effect, and for this reason it can be readily seen why most formulae are based on alcohol. The alcohol first used by Papanicolaou was ethanol, and in this country it is used in the form of 74 O.P. Industrial Spirit. Isobutanol and isopropanol have been used as alternatives to ethanol, but both reagents are more expensive than Industrial Spirit.

When used alone, Industrial Spirit is not a completely satisfactory routine fixative. Detail in some cells is blurred, making it difficult at times to define the limiting membrane of both the nucleus and the cytoplasm. Smears prepared in parallel with alcohol and another fixative show that a nucleus producing an irregular chromatin precipitate in the other fixative tends to "smooth out" when alcohol alone is used, thus giving a more benign appearance to the stained product; there is also a tendency for cell structures to take the basic stains more avidly.

Papanicolaou advocated alcohol - ether solution as the routine fixative of choice, and it is probably still the one most widely used. Almost all cytological fixatives have the disadvantage of being extremely volatile, and it should be noted that there is an extreme fire hazard when working with these solutions. For this reason containers supplied to collecting centres require an efficient seal; in this laboratory 100-ml glass jars with screw caps are used (see Fig. 1). The caps are fitted with replaceable waxboard liners.

Prolonged immersion of smears in any cytological fixative should be avoided. Changes in staining characteristics can be detected after about 24 h. If a delay greater than this is contemplated before processing, it is advisable to remove the slide from the fixative and allow to dry in air at room temperature. In practice, our peripheral collecting centres are instructed to decant the fixative and re-cap the jars before despatching them to the laboratory. Before staining, slides so treated should be re-immersed in the fixative for 15 min before proceeding to stain in the usual way.

Fixatives

(a) *Schaudin's solution*

    2 vols. of sat. aqueous mercuric chloride.
    1 vol. of 95% alcohol.
    Add 5 ml of glacial acetic acid to 95 ml of solution.

With this solution, as with all other mercury-containing fixatives, it is necessary to treat the films or sections to remove mercuric deposits before staining. The slides are placed in aqueous iodine followed by immersion in sodium thiosulphate solution (see ancillary Techniques—Rapid Paraffin Method, p. 119).

(b) *Alcohol - acetic acid*

    95% 74 O.P. alcohol.
    5% glacial acetic acid.

(c) *Formalin - alcohol*

    95% 74 O.P. alcohol.
    5% formalin (40% formaldehyde in water).

(d) *Alcohol - ether*

    50% 74 O.P. alcohol.
    50% diethyl ether.

(e) *Carnoy's fluid*
   30% chloroform.
   60% 74 O.P. alcohol.
   10% glacial acetic acid.

The optimum time for fixation of smears with any of solutions (a)–(e) is 20 min.

(f) *Methanol*
   100% methanol A.R.

This fixative is required for smears that are to be stained with a Romanowsky sequence. A fixation time of 10 min is an optimum. If smears are left immersed for longer periods, a background stain occurs, owing to the presence of a film of stained protein.

Commercial fixatives are available; examples are: a "drop-on" preparation "Cytofixative" (Ortho Pharmaceuticals Ltd); and an aerosol pack "Spraycyte" (Clay-Adams Ltd). A satisfactory British-made aerosol is "F.C.F." (Fountain's Aerosols Ltd, Bradford.)

Recovery of all the fixatives mentioned above may be practised, with the exception of Carnoy's fluid. In this case the complexity of assessing the proportionate losses of alcohol and chloroform renders the procedure uneconomical.

Used fixative, of the same formula, is emptied into stoppered Winchester or similar glass bottle containing a $\frac{1}{4}$-in layer of activated charcoal and anhydrous copper sulphate in equal parts. When the bottle is full it is thoroughly mixed and filtered through a double thickness of Whatman No. 1 paper. The filtrate is then brought to the correct specific gravity by adding the more volatile fraction of the fixative, usually alcohol, but ether in the case of alcohol - ether fixative.

A simple densitometer can be prepared by sealing a 5-ml glass ampoule loaded with lead shot. The stem is calibrated with a carefully prepared solution of the fixative to be recovered.

Aqueous fixatives may be re-used after simple filtration.

Dry Fixation

Recently attention has been drawn to the use of a "dry" fixation technique (Kohn and Earl, 1965). Slides are coated with the fixative and allowed to drain and dry. Smears are then prepared in the usual way and can be processed when dry.

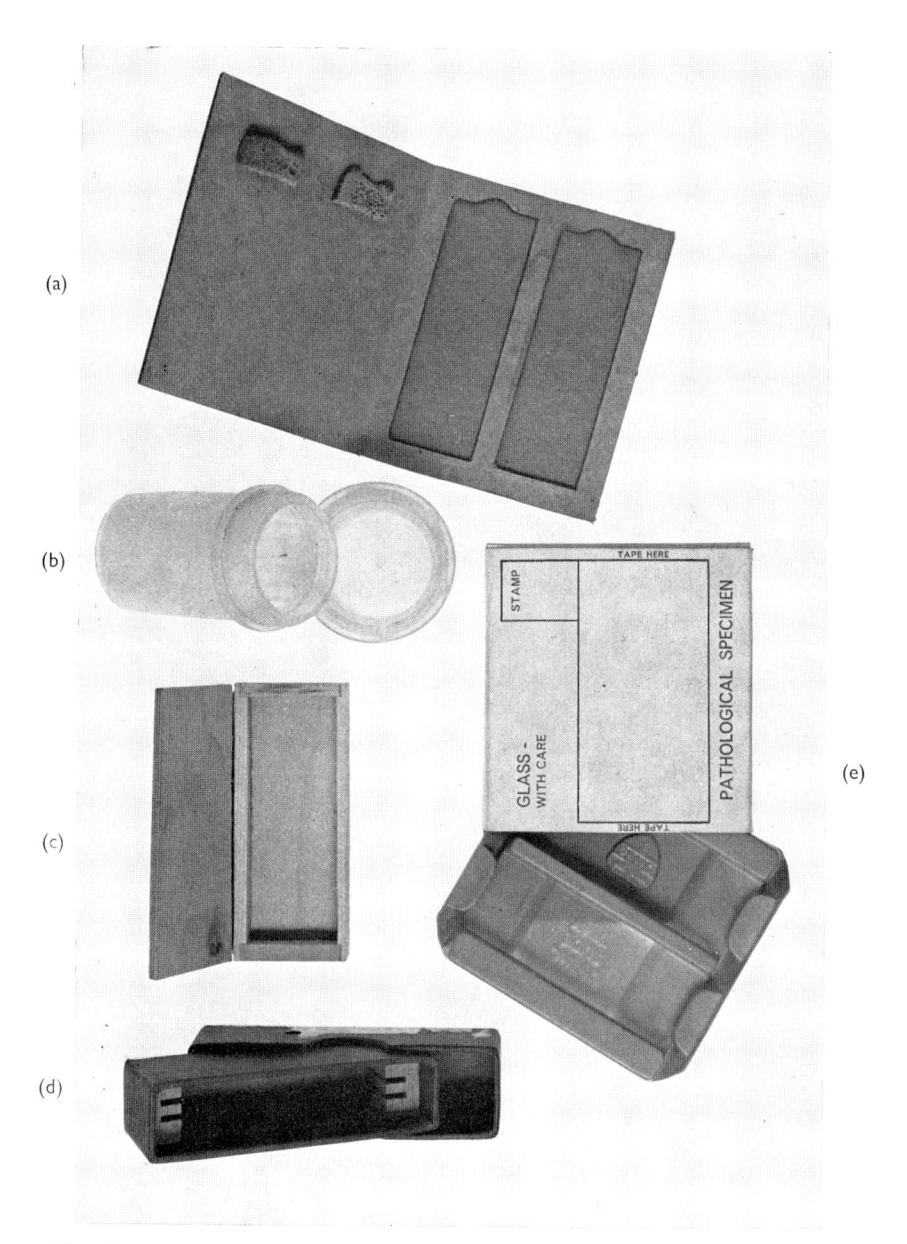

FIG. 2. Types of slide carrier for mailing smears: (a), paste-board wallet (Clay-Adams Ltd and Ames Ltd); (b), polythene jar (Wilson) (G. T. Gurr Ltd); (c), slide box, wood (Baird and Tatlock Ltd); (d), slide box, wood (A. Gallenkamp & Co. Ltd); (e), tray and sleeve (Ortho Pharmaceuticals Ltd).

Coating solution
    Ethanol, 30% v/v aqueous, 100 ml.
    Mercuric chloride, 5·0 g.
    Sodium acetate, 1·0 g.
    Sucrose, 5·0 g.

Drying Effect

The cells present in smears that have been allowed to dry before fixation show characteristic changes. Usually they are enlarged and show poorly defined structures, especially the borders of the nucleus and the cytoplasm. Chromatin loses its affinity for haematoxylin and the whole cell exhibits a weak response to basic stains.

Mailing Smears

Smears that have been properly prepared and adequately fixed can be safely removed from the fixative and allowed to dry before transit. Dry, unfixed smears are, generally speaking, quite unsatisfactory.

The greatest risks to which dry, fixed smears are exposed during postal delivery are contamination by dust, e.g., cotton fibres when cotton-wool is used as padding, and abrasion if the slides fit loosely in the container. These hazards can be largely eliminated by sheathing the slide in a polythene envelope before packing.

Slide boxes, of the conventional wood or paste-board type are satisfactory, provided the slide is firmly wedged. They are costly, however, in terms of capital outlay and stamp duty if large numbers are involved.

Two makes of stout cardboard wallet are available (Clay-Adams Ltd and Ames Ltd), both designed to hold two slides. As with the conventional slide box, the principal disadvantage is that the slides fit loosely in the depressions. This may be overcome by glueing small pads of foam rubber on the cover opposite the depressions (see Fig. 2a). A most useful pack consists of a plastic tray, holding two slides, that inserts into a cardboard sleeve. This extremely light, yet rigid, carrier is available from Ortho Pharmaceuticals Ltd. Illustrations of these carriers are shown in Fig. 2.

Cellophane strips can be placed in small Cellophane covers and may be despatched together with the request form, in a conventional paper envelope.

CHAPTER II

Staining, Mounting and Examining Smears

For the most part, routine cytology involves answering a request to exclude malignancy. This requirement is met by the use of haematoxylin, to stain nuclear structure, followed by a secondary stain that displays the remainder of the cell. Many cytology laboratories choose Papanicolaou's solutions for the secondary stain, although eosin is also used. Ability to stain nuclear structures is of paramount importance and pre-supposes proper fixation of satisfactory cells.

The choice of haematoxylin solution is a matter of personal preference, although Harris's formula is probably the most widely supported. It is necessary to use a regressive-staining technique whichever haematoxylin is selected. This involves allowing the solution to overstain the nuclei, followed by selective removal of excess of stain with a solvent—"differentiation".

A Romanowsky stain is a desirable ancillary technique, particularly for examining serous fluids, e.g., pleural and ascitic aspirations. The Jenner–Geimsa method described is not the simplest sequence of this type, but it is probably the easiest with which to achieve reproducibility. These stains are prepared from methylene blue, eosin and a series of related "azure stains". Broadly speaking, the methylene-blue components are the basic stains and the eosin components are the acid stains. Leishman's stain is perhaps the most generally used preparation in this class. However, when working with smears other than blood films, this stain can behave in a most capricious manner; for this reason the Jenner–Geimsa technique described is used in this laboratory. It must be remembered that methanol fixation is a pre-requisite for all Romanowsky sequences.

Jordan and Baker's modification of the Unna–Pappenheim stain is most helpful when decisions regarding the nucleic acids are needed. The stain differentiates between desoxyribonucleic acid (DNA) and ribonucleic acid (RNA), and its utility is further increased when used in conjunction with ribonuclease. Feulgen's stain is an additional aid in the assessment of nuclear DNA.

The sporadic excursions of cytologists, notably Bertalanffy and Bull (1960), into the field of fluorescence methods have yielded the acridine-orange method described (see Ancillary Techniques—Fluorescence Microscopy, p. 115).

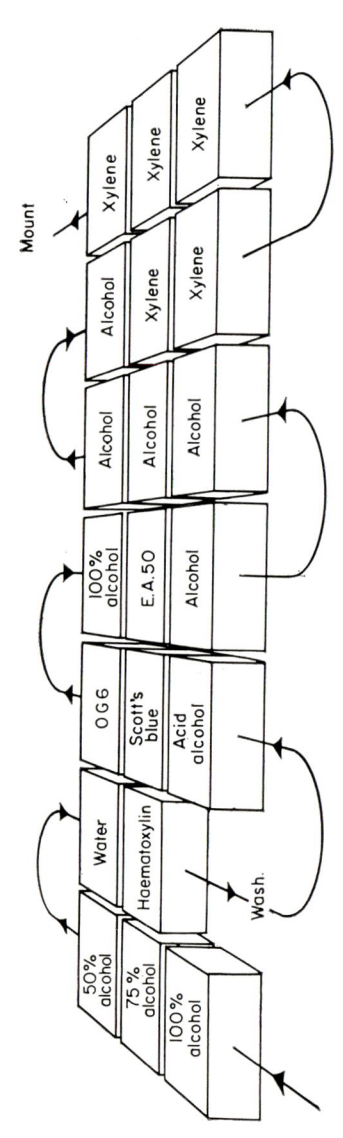

Fig. 3. Trough arrangement for the Papanicolaou staining sequence.

Staining Technique

Routine staining sequences, in which reasonably stable reagents are used, can be prepared in a series of glass staining troughs. A well fitting glass cover prevents concentration of the stain through loss by evaporation of the solvent. The troughs, together with slide carriers that hold ten slides, may be obtained from G. T. Gurr Ltd. Stainless-steel slide-carriers (Solmedia Ltd) that hold twenty-five slides and fit into the standard trough are available from Comber & Son, Heaton Mersey, Stockport. In this laboratory two sequences, Papanicolaou and Gram methods, are so prepared (see Figs. 3 and 4). The remaining stains, Feulgen, Jenner–Giemsa and Unna–Pappenheim, are prepared each day immediately before they are required and afterwards discarded.

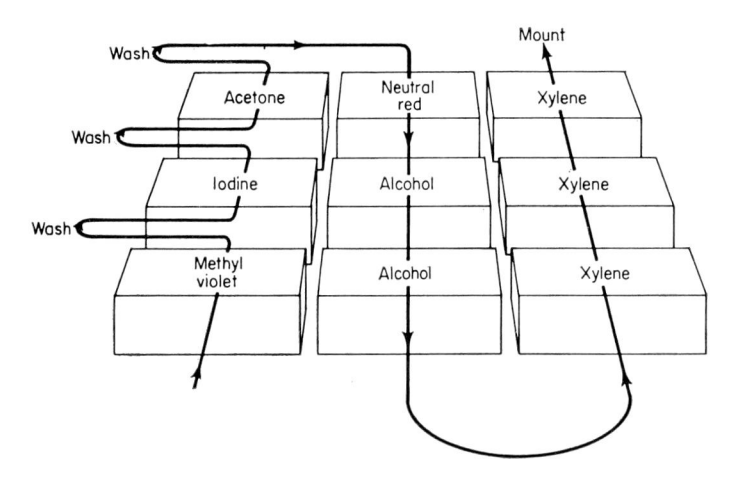

FIG. 4. Trough arrangement for the Gram staining sequence.

The length of time that the component staining solutions of the Gram and Papanicolaou methods may be used before renewal is a point for discussion. Experiment has shown, for example, that both sequences produce good preparations even after three months' daily use, provided care is taken to reduce "carry over" of the preceding solution to a minimum. This is accomplished by allowing the slides and carrier to drain for about 30 sec with an occasional gentle shake to remove the drops as they form. Holding the carrier at an angle helps this draining process. With the passage of around 80 slides/day, a weekly renewal of the staining solutions is quite adequate. Daily filtration through Whatman No. 1 paper keeps them free from debris. Frequent renewal of the graded alcohols and the xylene baths is necessary and is entirely dependent on the number of slides processed during any one day, but certainly daily changes are required.

Hydrating and Dehydrating Smears

The speed with which slides are passed through the alcohol and xylene baths of a staining sequence should be controlled to ensure that undue dilution does not occur. When the carrier is removed from a bath, the draining and shaking technique previously described should be used. On transfer into the next bath, the layer of solvent adhering to the slides and carrier must be equilibrated with the new solution. A visual control will accomplish this. When the slides are moved into the next bath, a rippling effect is seen across the slide if the carrier is moved in and out of the solution. The movement should be continued until this effect disappears before the carrier is passed on to the next bath.

Washing Smears

A golden rule to be observed in practically every staining sequence is to remove all the surplus stain or reagent from the smear before proceeding to the next step. This prevents the formation of "background stain" and, in cytology especially, anything that is likely to impair resolution of cell detail must be studiously avoided. During the immersion of smears in haematoxylin, for example, the stain is readily absorbed on to the film of mucoprotein that forms the background of the smear. If this excess of haematoxylin is not completely removed by washing, the subsequent brief immersion in acid alcohol during differentiation will be insufficient to accomplish this.

In most staining sequences, washing is carried out in running tap water; a few, e.g., Romanowsky and acridine orange, require buffered water, and still fewer, e.g., Papanicolaou's solutions, necessitate alcohol. When tap water is used, rubber tubing is led from the tap to the floor of the trough (see Fig. 5), which is then filled with water. The carrier is carefully lowered into the bath and a gentle stream of water allowed to flow. This creates a turbulence, which will carry surplus stain upwards and so leave with the overflow. Notice that the floor of the trough remains free of stain deposit. Allow the slides to wash for at least 3 min. An occasional gentle agitation will prevent areas of still water between slides.

If, during this process, fragments of the thicker parts of the smear become dislodged, careful examination should be made to ensure that the overflowing water is free from these particles before the carrier is removed from the bath, otherwise single cells, detached from the floating fragments may adhere to other slides.

It is unlikely that much information will be lost with these fragments, since their thickness precludes accurate cytology; however, it must be emphasized that the water should flow in a gentle stream.

Washing smears in buffered water rules out a continuous-flow technique.

For this kind of wash, a minimum of two baths is required. The first is used to remove gross excess of stain. This is achieved by dipping the carrier in the wash several times and gently agitating the slides during immersions. This process should occupy about a minute, and subsequent immersion with continual, gentle agitation in the second bath a further 2 min. The buffer solutions should be changed with each batch of slides by discarding the first solution and moving the second up to first place. The fresh solution then becomes the second wash.

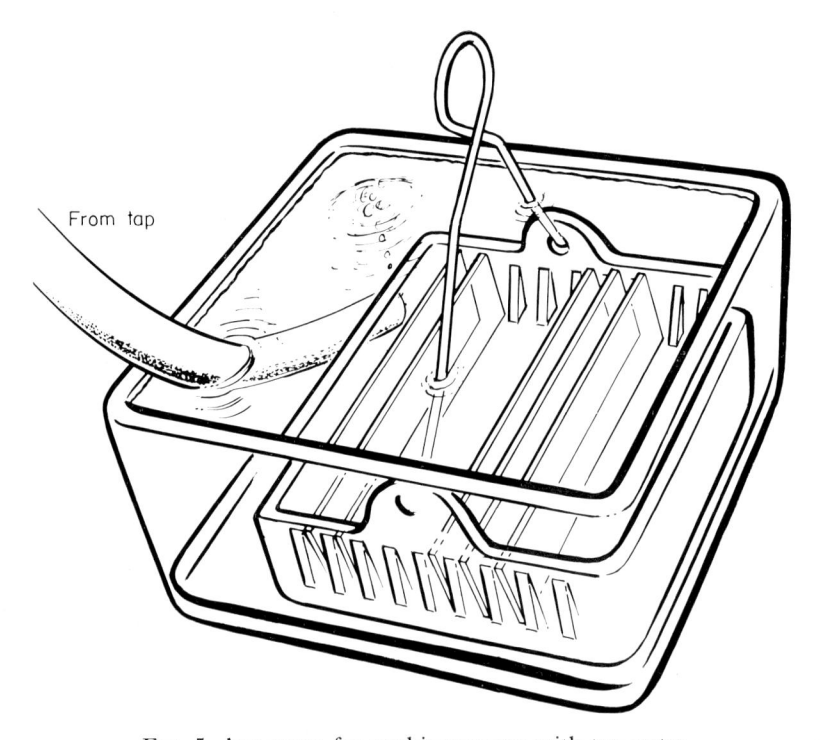

FIG. 5. Apparatus for washing smears with tap water.

Alcohol washes between stains require much shorter time intervals, and this information is given in the staining sequences involving this practice. During use, a series of alcohol washes quickly becomes contaminated with stain. When this becomes too obvious, the first bath is discarded and the others moved up one place; fresh alcohol is then introduced as the last bath in the series.

Similarly it is necessary to have more than one, preferably three, xylene baths. Periodically the first xylene is discarded and the remainder moved up to allow the fresh xylene to become the final bath. Slides should not be mounted from the first xylene, since this will often contain alcohol in sufficient concentration to interfere with the mountant.

Cleaning Smears

The first aqueous stage in a staining sequence presents an opportunity to clean the slides. The back is wiped with a laboratory tissue (Kleenex Medical Wipes are suitable) to remove any material inadvertently placed there during the preparation of the smear. Those extremities of the smear that will not be covered by the coverslip are wiped off by running the tissue carefully along the two edges of the slide. This leaves a clear margin about 1 mm wide down each side of the slide. If this step is not taken it is difficult to peel off mountant from the slides before they are filed. Two points to watch here are to use a clean tissue for each slide and not to allow the smear to dry during the cleaning.

Loading the Slide Carrier

Most carriers are designed to hold ten or twelve slides. An attempt to double the load with a "back to back" technique will often result in disaster: water trapped between slides is very difficult to remove. If extra slides are slanted across the carrier this may also lead to incomplete dehydration. It is safer, and usually quicker in the event, to use an extra carrier.

Differentiating

Smears are immersed in haematoxylin solution for a period such that the nuclei of the cells are just overstained. When such a smear is examined microscopically it will be seen that the nuclear sap of open nuclei (see The Normal Cell, p. 26) has taken up the stain. Differentiation is complete when the nuclear sap appears unstained. This is accomplished by immersing the slide in a solution of acid alcohol. It is not usually necessary to control individual smears. If the solution advised here is used, two consecutive immersions, each of precisely 1 sec, will yield practically 100% success. An occasional smear may require re-staining with individual treatment.

Blueing Smears

After differentiating haematoxylin-stained smears, the characteristic blue colour is changed to a nondescript red. Immersion in alkaline solutions causes an immediate reversion to blue. The quality of the blue colour is of primary importance in the examination of cell nuclei. Water that is "temporarily hard" contains calcium hydrogen carbonate, and is the best possible alkali to use for the process. Blueing can be carried out in the fashion described under "Washing" above. Laboratories not so fortunate in their water supply require a standard solution approximating to this. The formula given under "Scott's tap water substitute" is to be preferred.

Most alternatives, such as solutions of lithium carbonate and ammonium hydroxide, should be avoided, as these cause unnecessary loss of material from cytological smears.

Processing Cellophane Smears

The strip is cut into four pieces, each $1\frac{1}{2}$ in long. The first is left unstained, two are stained "Papanicolaou" and the fourth "Gram". Staining times are the same as for conventional smears on slides as far as the final xylene. The strip is coated by passing it through a 50% solution of standard DPX mountant in xylene by using forceps. It is then clipped to a rack and allowed to drain. These smears are safe to handle after 2 h at room temperature.

Re-staining

It is sometimes necessary to re-stain a smear either with the original stain or by another sequence.

The slide is placed in a Coplin jar containing xylene, covered and left overnight at 45°C. In almost all cases, the coverslip will have fallen from the slide by morning. Adhering mountant can be removed by passing the slide through three further jars of xylene, also at 45°C.

Re-hydration may then be carried out through alcohol, and the slide left to stand in 5% aqueous hydrochloric acid for 2 h. After this treatment it is freely irrigated with buffered water (pH 6·8) and left to stand for a further 2 h in the buffer, during which time several changes of the buffer solution are made. The smear is tested with BDH universal indicator to ensure that all the acid has been removed before proceeding to stain.

Conversion to Papanicolaou from Gram can be accomplished without preliminary decolorization, provided neutral red has been used as the secondary stain. After bringing the slide to alcohol as described above, it may be treated as if previously unstained. No obvious difference is seen in the final preparation.

Automatic Staining

A machine (Shandon Scientific Co. Ltd) (see Fig. 6) is available that successfully automates the Papanicolaou staining sequence. Personal deviations from the staining programme, which may occur in manual staining are thus eliminated. The consequent re-allocation of a technician's time together with the saving in bench space should be offset against the price of the machine (around £350).

Fig. 6. Apparatus for automatic staining.

Staining Sequences

The preparation of the reagents used in the following sequences is given under Formulae, p. 133.

In some laboratories the task of preparing routine-staining solutions is delegated to student or junior technical staff. It is, of course, important that students be instructed in the principles and practice of staining technique, and also that they be asked to prepare and subsequently use these solutions. Only in this way can the complexities be met and overcome. However, the practice ground for this learning should not lie in the staining routine of the laboratory. To ensure complete reproducibility, the solutions must conform rigidly to specification. If this precaution is observed and the staining times in a sequence are carefully followed, most of the black magic associated with stains disappears.

Compounding precisely formulated stains, such as those of Romanowsky, Papanicolaou, and the haematoxylins is an anachronism in present-day hospital laboratories. The armoury of technique covered by "quality control" in commercial preparations ensures a consistency that cannot be matched in the average pathology department without expending an inordinate amount of a skilled technician's time.

Haematoxylin and Papanicolaou

This is the most important sequence used in a cytology laboratory. The staining solutions are best obtained commercially prepared: preparations by Ortho Pharmaceuticals Ltd prove consistently satisfactory.

1. Alcohol (74 O.P.), 100% ⎫
2. Alcohol, 75%   ⎬ See note on hydrating smears, p. 10.
3. Alcohol, 50%   ⎭
4. Distilled water.
5. Harris's haematoxylin   4 min.
6. Wash in tap water   3 min.
7. Acid alcohol, 0·5%   2 sec. See note on differentiating, p. 12.
8. Wash in tap water   3 min.
9. Scott's solution   4 min.
10. Distilled water   Single rinse.
11. Alcohol, 50%   ⎫
12. Alcohol, 75%   ⎬ See note on dehydrating smears, p. 10.
13. Alcohol, 100% ⎭
14. O.G.6 solution   1 min.
15. Alcohol, 100%   Single rinse.
16. EA50 solution   1 min.
17. Alcohol, 100%   Single rinse.
18. Alcohol, 100%   Single rinse.
19. Alcohol, 100%   Single rinse.
20. Alcohol, 100%   Single rinse.
21. Xylene   Single rinse.
22. Xylene   Single rinse.
23. Xylene

Renew any bath that shows obvious contamination with fragments of material detached from smears. Change the Scott's blue (step 9), the water (step 10), the alcohol (step 17) and xylene (step 21) regularly (see Staining Technique, pp. 9 and 11).

The cell nuclei stain blue, the cytoplasm various shades of red (acidic or eosinophilic) or green (basic or cyanophilic).

Haematoxylin and Eosin

1. Alcohol, 100%
2. Alcohol, 75%
3. Alcohol, 50%

See note on hydrating smears, p. 10.

4. Distilled water.
5. Ehrlich's haematoxylin — 5 min.
6. Wash with tap water — 3 min.
7. Acid alchol, 0·5% — 5 sec.
8. Wash in tap water — 3 min.
9. Scott's solution — 4 min.
10. Distilled water — Single rinse.
11. Eosin — 2 min.
12. Distilled water — 30 sec.
13. Alcohol, 70%
14. Alcohol, 100%
15. Alcohol, 70%

See note on dehydrating smears, p. 10.

16. Xylene.
17. Xylene.
18. Xylene.

Cell nuclei blue, cytoplasm pink.

Weigert's Haematoxylin

This may be substituted in either of the preceding sequences at stage 5, with a staining time of 10 min and a 1% acid alcohol solution at stage 7 for 5 sec. Chromatin stains a blue-black rather than blue. This stain is sometimes used for re-staining if fixation is shown to be poor. Although it will improve the staining intensity of the nuclear chromatin of poorly fixed cells, it will not reduce the hazards of interpretation.

Shorr's Stain

1. Shorr's trichrome — 1 min.
2. Alcohol, 100%
3. Alcohol, 100%
4. Alcohol, 100%

See below.

5. Xylene.
6. Xylene.
7. Xylene.

The smears are moved directly from the fixative into the stain. Two seconds in each alcohol will produce well differentiated smears. It is not necessary to use a graded series of alcohols, since the slide has not been exposed to water.

This stain differentiates between cornified cells (which stain red) and uncornified cells (which stain green), and is used in hormonal work.

Gram Stain

1.	Methyl violet solution	2 min.
2.	Wash in tap water	3 min.
3.	Aqueous iodine	2 min.
4.	Wash in tap water	3 min.
5.	Acetone	See below.
6.	Wash in tap water.	
7.	Neutral red solution	5 min.
8.	Distilled water.	Rinse
9.	Alcohol, 100%.	3 sec.
10.	Alcohol, 100%.	2 sec.
11.	Alcohol, 100%.	Single rinse
12.	Alcohol, 100%.	Single rinse
13.	Xylene.	
14.	Xylene.	
15.	Xylene.	

Slides can be passed directly from fixative into the methyl violet. The iodine solution begins to weaken after daily use for a week, when it should be changed.

Do not keep the acetone bath for longer than 1 day. A reaction that occurs during differentiation produces a volatile irritant.

The first rapid immersion in acetone, duration 1 sec, removes all the loosely adherent methyl violet that has resisted the tap-water wash. A second, briefer, immersion of the carrier in the acetone bath causes unbound methyl violet to be leached from the smears. This step is immediately followed by transfer of the slides into tap water, where they are freely irrigated. If this differentiating technique is followed, slides can be safely processed in batches, although thick areas of smears may be under-differentiated.

Biebrich Scarlet — Fast Green (Guard's Stain)

1.	Alcohol, 70%	2 min.
2.	Biebrich scarlet solution	2 min.
3.	Alcohol, 50%	Rinse.
4.	Fast green solution	See below.
5.	Alcohol, 50%.	5 min.
6.	Alcohol, 80%.	2 min.
7.	Alcohol, 100%.	1 min.

8. Alcohol, 100%.   1 min.
9. Xylene.
10. Xylene.

The differentiation at step 4 must be controlled by microscopy at 30-min intervals after the first hour. This usually takes about 3 h, and is complete when squamous cells with vesicular (open) nuclei are stained green. Pyknotic nuclei and the sex chromatin body remain red.

Cresyl Fast Violet

1. Alcohol, 100%.   ⎫
2. Alcohol, 75%.   ⎬ See note on hydrating smears, p. 10.
3. Alcohol, 50%.   ⎭
4. Distilled water.
5. Cresyl fast violet solution   10 min.
6. 0·01N acetic acid   Wash.
7. 0·01N acetic acid   Wash.
8. Distilled water   Rinse.
9. Alcohol, 100%   Rinse.
10. Alcohol, 100%   See below.
11. Xylene.
12. Xylene.

The sex chromatin is seen in vesicular nuclei as a deeply stained ovoid body contiguous with the nuclear membrane (see Nuclear Sex, p. 131). Differentiation, step 10, is controlled by microscopy until nuclear detail can be clearly seen.

Leishman's Stain

1. Staining solution   4 min.
2. Buffer solution, pH 6·8   Wash.
3. Buffer solution, pH 6·8   Differentiate—see below.
4. Allow to drain and dry in air.
5. Xylene.   5 min.
6. Xylene.

The staining solution is prepared immediately before use by diluting one volume of the stock stain with two volumes of buffer solution (pH 6·8).

Differentiation is complete when the nuclear chromatin is blue and the cytoplasm a transparent pale blue. Granules present in the cytoplasm stain in a range of colours from blue back to red and help to characterize particular cell types. In the presence of large numbers of red cells, this differentiation may be controlled by naked eye. At the point when the appearance of the film changes to a deep pink colour, differentiation is complete. Coincident with this is a sudden lightening in the blue colour of the buffer solution running off the slide when it is lifted from the wash. This may be clearly seen if the wash from the slide is allowed to fall on to clean blotting-paper. Smears containing few red cells can be differentiated without recourse to microscopy by using this change in the colour of the wash as a guide.

Jenner–Giemsa Stain

1. Dilute Jenner stain   2 min.
2. Dilute Giemsa stain   4 min.
3. Buffer solution   Wash.
4. Buffer solution   Differentiate—see below.
5. Drain and dry in air.
6. Xylene.   5 min.
7. Xylene.

The staining solutions are prepared freshly each day by diluting stock Jenner with two volumes of buffer and stock Giemsa with eight volumes of the buffer (pH 6·8).

Slides are moved directly from the Jenner into the Giemsa without rinsing.

Differentiation is accomplished in the same way as for Leishman's stain, except that the presence of large numbers of red cells in the smears produces a reddish-purple rather than a pink colour.

The 5-min immersion in the first xylene bath ensures that all the air is removed from the surface of the film before mounting. If water is still present on the slide incident light is almost wholly reflected, producing a mirror effect on immersion in xylene.

Feulgen's Stain

1. Alcohol, 95%.
2. N hydrochloric acid   1 min.
3. N hydrochloric acid, at
     60°C   20 min.
4. N hydrochloric acid, cold   1 min.
5. Distilled water   Rinse.
6. Feulgen's solution   2 h.
7. Sulphurous acid   5 min.
8. Sulphurous acid   10 min.
9. Sulphurous acid   5 min.

10. Distilled water   Rinse.
11. Distilled water   Rinse.
12. Drain and dry in air.
13. Tartrazine solution   1 min.
14. Xylene.
15. Xylene.
16. Xylene.

The sulphurous acid is freshly prepared just before it is required. Solutions from step 2 to 11 inclusive are discarded each day.

DNA present in the cells is stained purple; all other structures are stained yellow.

Unna–Pappenheim Stain

1. Alcohol, 100%.
2. Alcohol, 75%.
3. Alcohol, 50%.   } See note on hydrating smears, p. 10.
4. Distilled water.
5. Unna–Pappenheim stain   20 min.
6. Distilled water   Rinse quickly and blot dry.
7. Acetone   Rinse quickly.
8. Acetone - xylene, equal
   volumes   Rinse quickly.
9. Xylene.
10. Xylene.
11. Xylene.

The stain differentiates DNA from RNA, the DNA staining blue-green and the RNA red. High concentration of RNA cause the cytoplasm to stain an intense red. This may be confirmed by exposing a duplicate slide to a 0.1% solution of crystalline ribonuclease in water for 1 h at 37°C. Subsequently this slide is stained in parallel with an untreated slide.

Mounting Smears

With conventional methods, mounting large numbers of slides is time consuming. Some of this time can be saved by short-cuts that have proved safe in our routine processing. The coverslips we use are Chance glass Grade No. 1, one of five thicknesses available as follows:

Grade No.	Thickness (mm)
0	0·085–0·130
1	0·130–0·160
1½	0·160–0·190
2	0·190–0·250
3	0·250–0·350

Coverslips of greater thickness than No. 1 impair the resolution of nuclear detail and may prevent correct focusing of objectives having short working distances. There is no need to immerse the coverslips in a cleaning fluid before use, and, if carefully handled, contamination from the fingers will not occur. A suitable size of coverglass is 22×50 mm: this will cover an area of the slide, leaving just sufficient space to carry the standard slide label. The self-adhesive type of label is superior to those backed with a water-based gum. After several years' storage, the gummed labels quite often become detached.

As with stains, the mountant is best obtained as a commercially prepared solution. A polystyrene type, such as DPX, is the most satisfactory. Transparency is not lost with "ageing", as so often occurs with balsams, and the plasticizers in commercial preparations prevent shrinkage.

For mounting the slides, a large sheet (46×57 cm) of Whatman No. 1 paper is laid out flat on the bench—a Formica-top bench is ideal. This will hold six dozen cover glasses with ample room between them. The coverslips are first tipped straight from the box on to a second sheet of fluffless paper. The lower surface is cleaned by drawing the coverslip over the paper towards the edge of the bench with the finger tip. It is then carefully lifted so that its under surface becomes the face and placed face upwards on the first sheet of paper. When a sufficient number have been so treated, DPX is dropped on to the coverslips by using a squeeze-controlled polythene dispenser of the type used for washing-up detergent (see Fig. 7).

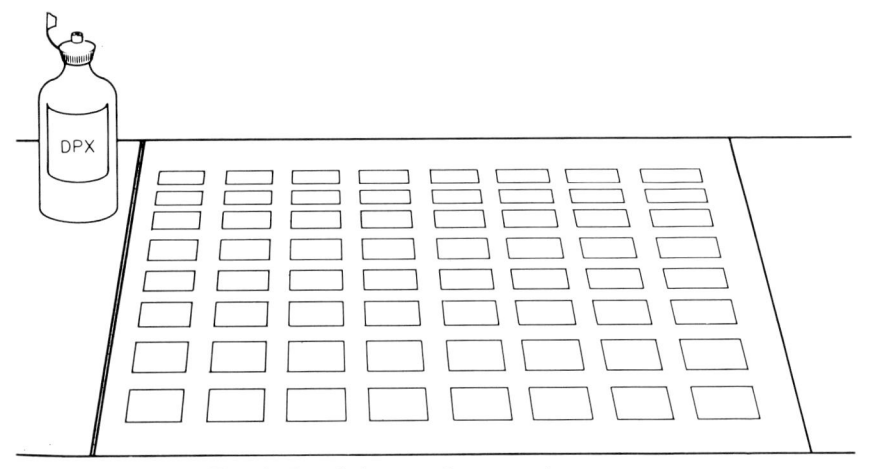

FIG. 7. Bench lay-out for mounting smears.

The slides to be mounted are removed from the last xylene bath one at a time. Each is placed smear downwards on to a coverslip and firm pressure applied to drive the excess of mountant to the edges of the sandwich. It is important to see that the xylene has not evaporated from the smear before contact with the DPX is made. Practice quickly shows the amount of DPX to place on the coverslips and the time elapsing between the addition of the DPX and the mounting is sufficient to allow the DPX to spread evenly over the coverslip. The mounted slides are lifted from the paper and allowed to dry out in slide trays overnight. Excess of mountant is peeled off the following morning when the slides are labelled. A wipe with a soft cloth removes finger-prints on the coverslip. If Canada balsam mountant is used, a practice advocated by Graham (1963) is to place the freshly mounted slide on a hot plate and to apply pressure with forceps on the coverslip until the xylene vapour has been driven out. The slides are then left to cool, after which they are ready to use.

Varnishes have been used, which eliminate the necessity of the coverslip. An aerosol preparation suitable for cytological work (Shaw, 1965), is Artists Picture Varnish (Windsor & Newton Ltd). The slides are transferred from the xylene bath to slide trays; a paper mask is used so that only the smear area is exposed. When most of the xylene has evaporated, the varnish is sprayed evenly over the smears. Drying occurs within a few minutes, and the varnish is completely set in about half an hour. Oil-immersion objectives can be used, but subsequent removal of the oil should be accomplished with a tissue; if xylene is applied, the smear will require re-varnishing.

Permanent mounts of Cellophane smears are prepared by sandwiching the strip between slide and coverslip with DPX. Temporary mounts are prepared by substituting immersion oil; these strips can be recovered by removing the coverslip and wiping the Cellophane free from oil. The slides and coverslips may be safely re-used after washing in xylene.

A blue background is present in Cellophane preparations stained with haematoxylin. This is not intense and does not affect critical assessment of the cells.

Examination of Smears

Microscope

A suitable microscope is a binocular instrument with a quadruple nosepiece, the lenses giving final magnifications of $\times 50$, $\times 100$, $\times 400$ and $\times 1000$. If a quintuple nosepiece is fitted, a dry lens giving a magnification $\times 500$ is a useful addition. When the choice of objectives lies between a dry lens and an oil-immersion system, the dry lens should be selected. Scanning continuity is lost when a change of objectives involves the addition or removal of immersion oil. The light source should be integrated with the instrument, and a control that varies the intensity of illumination is most important. Additional equipment, filters and a microscope hood or dark room is necessary for fluorescence techniques.

In this laboratory, microscopes manufactured by Gillet and Sibert have been found entirely satisfactory and relatively inexpensive multi-purpose instruments. The "Conference" model can be used for normal bright-field microscopy, dark-ground and fluorescence techniques, group viewing, projection and photomicrography. It can be purchased as a basic "bright-field" apparatus and the various accessories added as they become necessary. To perform all these functions, with the exception of photomicrography, the cost will be around £300. A most important factor and economy in the specification is that the light source, a 12 V, 100-watt tungsten iodide lamp, is used for all the techniques.

(a) *Positioning*

Although many textbooks advocate placing the microscope so that windows passing direct light are behind the worker, a preference is sometimes shown for using the instrument directly in front of a window. The prolonged microscopy and the intensive concentration associated with cytological screening demands the best possible position and environment. A room of such a size as to accommodate only one worker comfortably and possessing adequate window space is the ideal. The microscope bench should extend beyond the window width with the microscope facing a wall (see Fig. 8). This arrangement allows long periods of uninterrupted concentration and at the same time gives access to a distant horizon.

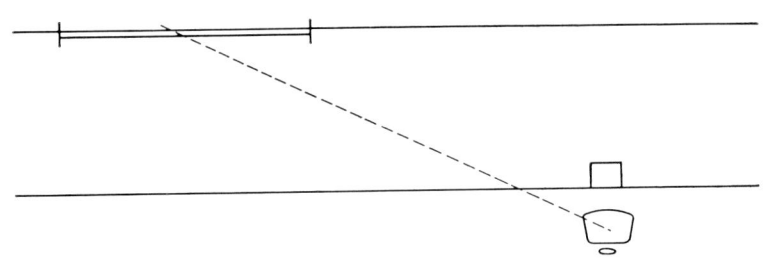

Fig. 8. Diagram showing position of the microscope in relation to the window.

(b) *Intensity of illumination*

Initially in training it will be found necessary to work with a brighter field than is usual in routine microscopy. Once the initial training period has passed, it becomes possible to examine smears without stepping up the illumination on more than one or two occasions. A headache is the reward for long periods of continuous bright-field microscopy in the early stages. For the trained worker, critical assessment begins to fall off after the second hour at the microscope.

(c) *Scanning technique*

Before commencing each microscope session, the routine "setting-up" procedure recommended by the manufacturers should be carried out;

particular attention should be given to cleaning the lenses. One should never assume that the instrument is ready for use.

The slide is first examined with a low-power objective (×10) in order to cover the whole of the smear. This is an essential preliminary step that allows the examiner to enlarge on information gained from the macroscopic assessment. During this survey, a knowledge of the topography of the smear is obtained. The disposition of plaques of cells and fragments of tissue is noted, because atypical cells can often be related to these aggregates. As experience accumulates, this preliminary examination will yield sufficient information to allow the worker to form a definite opinion concerning the cells in the smear. Scrutiny of selected areas with a stronger objective (×40) is usually sufficient to confirm the impression.

During the training period in cytological screening, however, re-examination of the entire smear with the ×40 objective should be meticulously carried out. This is a time-consuming process, but, even so, every cell must be examined and classified if possible. Any abnormal cells are noted and an explanation sought. If it is found difficult to identify a cell, its location should be marked and experienced opinion obtained.

Marking Cells

It is frequently necessary to be able to return to particular cells that have been seen in smears, either because they are "reference cells", showing individual characteristics, or because another opinion is required. Orthodox microscopy involves the use of the verniers attached to the microscope stage for locating these cells, but, on occasions, this method may result in wasted time. A substitute technique, widely practised in cytology, is to place an ink mark on the coverslip. "Reference marks" must be durable, since the slide will be in use for several years. Gurr's black glass-writing ink and a fine drawing pen serve this purpose. The pen is washed free from ink after use by dipping it into a jar of water; excess of water is shaken off and the nib allowed to dry in a rack on the bench. An American company, Zieler Instrument Co., Massachusetts, manufacture an apparatus that can be screwed into the nosepiece of the microscope. To mark a field, the marker is swung into the optical axis and the outer sleeve pulled down towards the coverslip: on contact a circle, concentric with the optical axis, is stamped on the smear.

Standards for Cell Size

Critical measurement of cells is not especially helpful in routine examinations. A rough estimate of cell size can be made by comparison with cells that have a fairly constant morphology and that are usually present in the smears. Two examples are the erythrocyte (about 7μ in diameter) and the neutrophil (about 12μ in diameter).

CHAPTER III

Structure of the Cell

The General Cell

No single animal cell can be regarded as completely unspecialized. The notion of a general cell, however, affords a basis for the understanding of characteristics common to many: its shape, when free-living and resting, would be spherical (see Fig. 9).

The contents are contained within a thin, limiting membrane or plasmolemma. This is a cohesive film of protein molecules that forms a selective barrier between the cell interior and the fluid outside. Generally, water with dissolved crystalloids and gases—chiefly oxygen and carbon dioxide—can traverse the barrier, but not proteins. Within the cell, two major divisions of the contents are easily seen: the background of finely granular cytoplasm and a more dense spherical structure, the nucleus. A second membrane, the nuclear membrane, separates the cytoplasm from the nucleus.

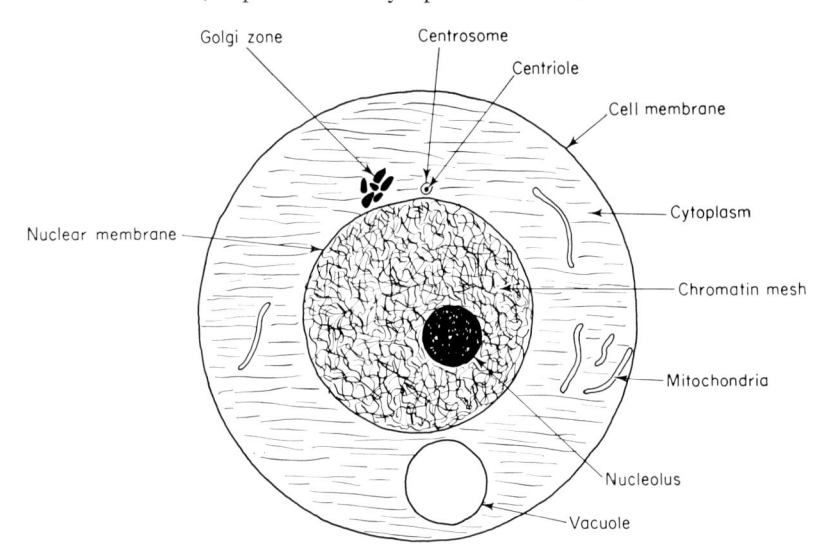

Fig. 9. Diagram of the general cell.

The nuclear substance can be resolved into three principal parts:
1. A fine meshwork of protein fibres with strongly basophilic staining properties that, because of this characteristic, has been named chromatin.

2. A spherical structure that stains differently from the chromatin, the nucleolus.
3. A weakly staining background of nuclear sap.

Within the cytoplasm, close to the nucleus, can be seen the centrosome. This contains the intensely staining centriole that takes part in cell division. Associated with this is an area of refractile fat globules, the Golgi zone. Small rod-shaped bodies may also be seen: these are the mitochondria that are concerned with the metabolism of the cell. Vacuoles, if present, are seen as unstained, spherical areas in the cytoplasm and indicate food assimilation, excretion or secretion by the cell. Inclusions may also be present in the cytoplasm in the form of irregular granules. These may be due to pigment or ingested foreign material.

The increased magnification available with electron microscopy yields much more information concerning cell structure, but these details are of no practical value in routine clinical cytology.

Cell Reproduction

Mitosis

Cells that reproduce themselves within a tissue in order to maintain the life and form of an organism are referred to as somatic cells. Changes occur in the chromatin when a cell is about to divide.

The meshwork of nucleoprotein filaments, seen in the resting nucleus, disentangles and, as division progresses, the constituent threads coil and shorten, becoming recognizable as chromosomes. The total number and morphology of the chromosomes—the chromosome complement—is normally constant for a particular species and comprises pairs of chromosomes, one of each pair deriving originally from each parent. This constant is called the diploid number of the species, and is given the cypher $2n$. In man, the chromosome complement is 23 pairs giving the diploid number 46. Constancy in the diploid number is associated with exact replication of the parent cell. This is accomplished by the division of each chromosome along its long axis into two halves. Each of two daughter nuclei produced contains half of every chromosome. This process of cell replication is called mitosis.

Meiosis

Fertilization, the fusion of male and female germ cells (gametes), achieves an intermixing of chromosomes. The diploid number of a species is preserved by a halving process during gametogenesis. These cells, therefore, contain nuclei that possess half the diploid number of chromosomes. They are termed haploid nuclei, and are given the cypher n.

After fertilization, the resulting cell (zygote) will possess a diploid ($2n$)

nucleus, since each gamete supplies n chromosomes. In man, each gamete supplies twenty-three chromosomes. Forty-four of the chromosomes, twenty-two from each gamete, can be matched to form homologous pairs, and are called autosomes. The remaining "odd pair", one from each gamete, represents the sex chromosomes. This cycle of events, which results in the production by a diploid ($2n$) cell of cells with haploid (n) nuclei, is called meiosis.

The Chromosome

Chromosomes are threads of nucleoprotein that, in the resting nucleus, appear as a fine reticulum: profound changes occur with the onset of mitosis.

It is thought that at prophase (see The Mitotic Cycle, p. 29), nucleic acid from the nucleolus coats the thread, which then coils into a tight spiral. When the fibre uncoils, during telophase, the nucleic acid is released and the nucleolus re-forms. There are two main kinds of nucleic acid that differ in the type of sugar, a pentose, and the organic bases present in them. These nucleic acids are synthesized from polymerized nucleotides, which have the general composition pentose+phosphoric acid+four organic bases:

	Nucleotide constituting DNA	Nucleotide constituting RNA
Pentose	Desoxyribose	Ribose
Acid	Phosphoric	Phosphoric
Bases	Adenine, guanine, thymine, cytosine	Adenine, guanine, uracil, cytosine
Occurrence	Present round the chromosome	Present in the nucleolus and the cytoplasm. (Coats the chromosome during division)

In general structure, a chromosome consists of two threads of nucleoprotein, called chromatids, which are joined at a single point, the centromere (Fig. 10). During certain stages of cell division, beads of

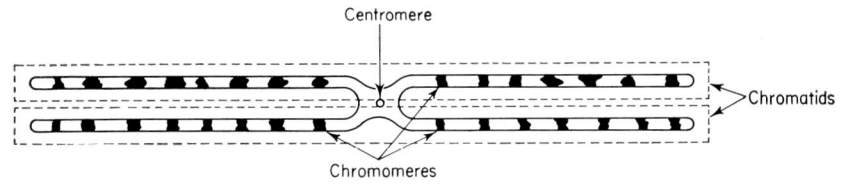

FIG. 10. General diagrammatic structure of a chromosome.

darker-staining protein may be seen along the arms of the chromatids. These beads are called chromomeres and the lighter areas interchromomeres. Chromatids carry the genes that determine hereditary characteristics; the centromeres are genetically inert.

Careful examination of the normal human somatic nucleus, at a stage in mitosis called metaphase, shows differences in morphology to exist amongst the chromosomes (Fig. 11). After matching the chromosomes into homologous pairs these differences can be used as a means of classification, the clinical application of which is referred to later (see Ancillary Technique—Tissue Culture, p. 125). In 1960, a meeting of cyto-geneticists at Denver, Colorado, formulated standards for classing human chromosomes into seven groups of autosomes, plus the two sex chromosomes. This classification is known as the Denver system and is a generally accepted standard method.

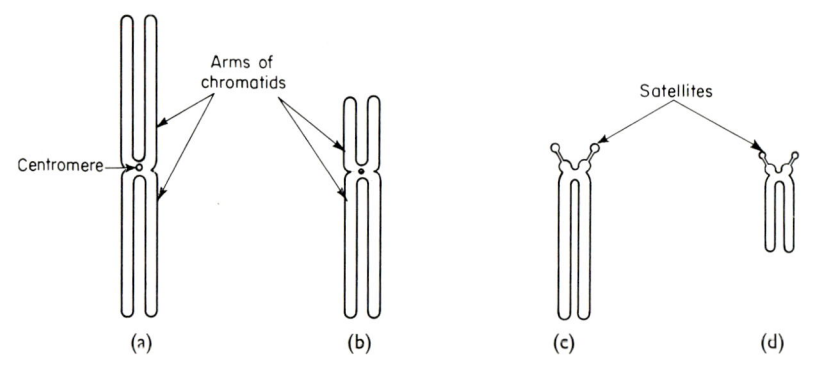

FIG. 11. Types of human chromosomes: (a), median centromere (arms of chromatids equal); (b), sub-median centromere (arms of chromatids unequal); (c) and (d) acrocentric centromeres, (c) long arms, (d) short arms.

The criteria are:
1. The overall length of the chromosome.
2. The position of the centromere. When the centromere is equidistant from the ends of the chromosome, or almost so, it is termed a median or metacentric chromosome. If the centromere is subterminal, the chromosome is described as acrocentric, and if the centromere is located between these positions it is called a submedian or sub-metacentric chromosome.
3. The lengths of the arms of the chromatid.
4. The presence or absence of satellites (this term is used to describe the short arm of a chromatid when it has a marked constriction at a point near the centromere). This constriction gives the appearance of the distal portion of the chromatid "budding off".

The first two criteria serve to arrange the pairs of chromosomes into seven groups. Often it is difficult to orientate the chromosomes within a particular group.

The groups are:

Group	I	Chromosome pairs 1, 2 & 3	The largest, metacentric chromosomes.
Group	II	Chromosome pairs 4 & 5	The next largest, sub-metacentric chromosomes.
Group	III	Chromosome pairs 6, 7, 8, 9 10, 11 & 12	All sub-metacentric.
Group	IV	Chromosome pairs 13, 14 & 15	Acrocentric with satellites.
Group	V	Chromosome pairs 16, 17 & 18	16 metacentric, 17 and 18 sub-metacentric.
Group	VI	Chromosome pairs 19 & 20	Metacentric.
Group	VII	Chromosome pairs 21 & 22	Acrocentric with satellites.

Identification of the sex chromosome may cause difficulty; the X chromosome resembles the autosomes in Group III, and the Y chromosome those of Group VII. During the classification, therefore, the sex chromosomes are aligned at the end of the autosome group they resemble (see Fig. 12). A subsidiary examination, the determination of chromosomal sex helps to resolve this problem (see Ancillary Technique—Chromosomal Sex, p. 131). A normal chromosomal female nucleus will contain two X chromosomes, and a normal male nucleus will contain one X and one Y chromosome.

Karyotype

The number and morphology of the constituent chromosomes of a particular cell is called its karyotype and, in this work, infers their classification into groups as described. The total number of chromosomes present is referred to as a *chromosome count*.

The *chromosome number* refers to the number of chromosomes per nucleus considered to characterize an individual and is decided after chromosome counts have been made on several (30 to 50) cells. When this has been carried out, a diagram, indicating the considered chromosome complement of the average nucleus, is drawn (see Fig. 12). This is termed the chromosome idiogram of the individual.

The Mitotic Cycle

The sequence of events followed by the replicating somatic cell is called the mitotic cycle. Although it is continuous it can be roughly divided into

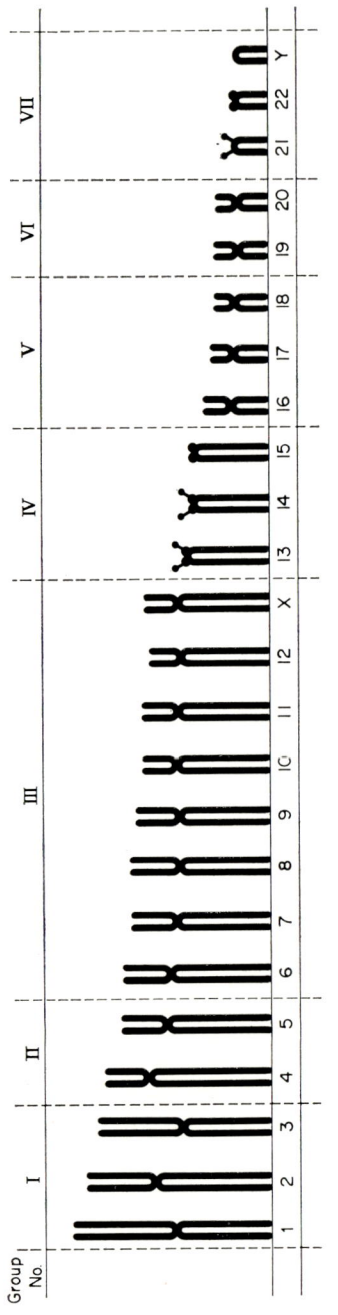

Fig. 12. Idiogram of chromosomes of male nucleus.

STRUCTURE OF THE CELL 31

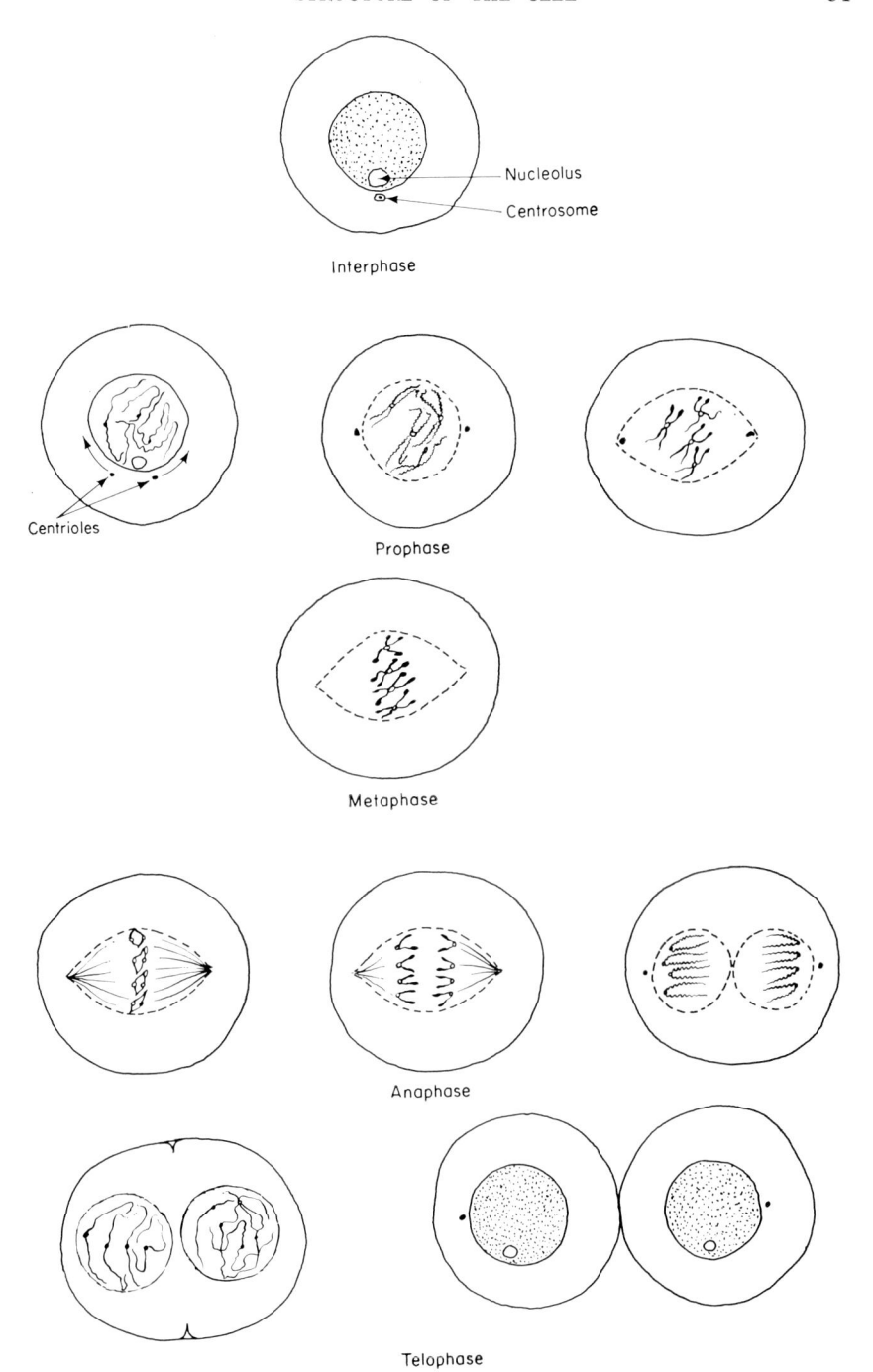

Fig. 13. Stages in mitosis.

five recognizable stages by conventional microscope techniques (see Fig. 13). It must not be assumed, however, that the cycle is divided equally in time between these stages. The longest interval is interphase, occupying hours or days, whereas anaphase may be completed in seconds or minutes.

The five stages are named: prophase; metaphase; anaphase; telophase; and interphase (see Fig. 14).

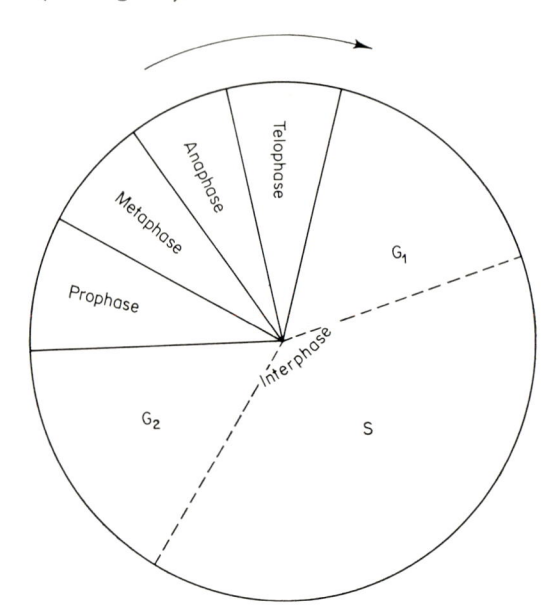

FIG. 14. The mitotic cycle.

Prophase

The centrosome breaks down, releasing the divided centriole. These migrate to form the mitotic asters, which are the diametrically positioned focal points of the daughter nuclei. The nucleolus disappears while the chromosomes disentangle, the fibres of nucleoprotein coiling on themselves forming tight spirals and thereby becoming much shorter, thicker structures. This process is described as condensation by helical coiling. At this point the nuclear membrane breaks down, marking the end of this stage.

Metaphase

Disintegration of the nuclear membrane brings the contents of the nucleus in contact with the cytoplasm. The nuclear sap precipitates as a fibrous gel to form a lemon-shaped structure, called the spindle. The extremities of the axis of the spindle, called the poles, are formed by the asters released from the centrosome. During the formation of the spindle, the chromosomes continue to coil and become progressively shorter. They move with a continuous oscillatory motion until, at a particular instant, the centromere of each chromosome becomes aligned at the equator of the spindle—the phragmoplast. Recent study suggests that each chromatid is attached to the spindle by a "centric fibre", which exerts a pulling force on the centromere. This marks the end of metaphase.

Anaphase

At this point the chromosomes contract to a maximum. The spindle stretches and the chromatids move to opposite poles. This marks the end of anaphase.

Telophase

The chromatids, having moved to the apices of the spindle, uncoil and division of the cytoplasm occurs by a cutting inwards along the phragmoplast. The nuclear membrane and the nucleolus reform, and the characteristic contracted structure of the chromatids disappears.

Interphase

Investigation has shown that this stage, when the daughter nuclei build complementary chromatids, is divided into three periods: a pre-synthetic period, called G1; a synthetic period, called S; and a post-synthetic period, called G2. These phases, however, cannot be distinguished with routine microscopy techniques. The G2 phase of interphase is synonymous with the "resting nucleus", with its fine, regular reticulum of nucleoprotein. Nuclei in this state are said to have an "open" or "vesicular" chromatin pattern.

Pyknosis

Nuclei of adult cells that retain an open reticular pattern are usually functional throughout the life of the cell: those that pyknose are not. Pyknosis infers that nuclear function has been fulfilled and the adult phase of the cell's life has begun. The young erythrocyte, for example, requires a functional nucleus in order to synthesize haemoglobin. When this process is complete, the nucleus undergoes progressive condensation (pyknosis) until it finally fragments (karyorrhexis) and is extruded from the cell. Not all normal cells whose nuclei undergo pyknosis are anucleate when adult. The superficial layer of some stratified epithelia, for example the genital and respiratory tracts, consists of squamous cells whose nuclei, having participated in the production of an exoskeleton of scleroprotein and interstitial cementing substance, undergo condensation but not karyorrhexis. They remain in the cell as pyknotic nuclei.

Meiosis

Although meiosis is a continuous event, two distinct phases occur. The first is the reduction process followed immediately by a mitotic division.

The chromosomes appear as long, thin, single strands, not split into chromatids. Homologous chromosomes associate in pairs so that the gene positions of one correspond with the gene positions of its fellow. Pairing begins at the centromere or the ends of the thread and proceed in zipper-like fashion. Each chromosome of the pair divides into chromatids, so that there are now four chromatids lying side by side. Exchange of material occurs during this division of chromatids: the points of breaking are termed chiasmata, and the exchange of linked fragments is termed "crossing over". The chromosome pairs shorten and become arranged on the equator of the spindle similar to mitosis. The difference is that homologous chromosomes pass to the apices and the daughter masses now possess half the chromosome number. The second phase is a mitotic division that follows immediately without reconstitution of the nucleus, thus yielding four haploid cells.

For a fuller introduction to the mechanics of meiosis the reader is referred to *Zoology for Intermediate Students* (Chapman and Barker, 1964).

Epithelial Tissues

An epithelium is a tissue that forms the limiting surface of other body tissues; the skin and the cell lining of blood vessels are epithelia. The constituent cells of an epithelium are called epithelial cells.

After fertilization, the diploid cell produced undergoes successive divisions leading to the formation of three layers of cells. The different layers are called ectoderm, mesoderm and endoderm. In the developing embryo, these germinal layers proliferate and differentiate to produce the specialized tissues of the systems and organs of the body. All three layers produce epithelial tissues, and it may not be possible to distinguish between epithelial cells arising from different germinal layers on morphological grounds. The flat, polyhedral squamous cell, for example, can be produced by all three layers.

The ectoderm produces the skin and nervous system, including parts of the sense organs.

Mesoderm gives rise to the vascular system (including the heart), the skeletal system (including muscle) and other supporting tissues. Part of the mesoderm retains its epithelial character and produces the squamous epithelial lining of the body cavities, i.e., the lining of the peritoneal, pericardial and pleural spaces. This type of epithelium is referred to as mesothelium. Two other types of epithelia are produced by the mesoderm: endothelium and mesenchymal epithelium. Endothelial cells form the single layer of squamous cells lining the vascular system and lymphatics. The squamous epithelial cells lining the subdural and subarachnoid spaces constitute the mesenchymal epithelium.

The endoderm supplies the epithelial lining and glands of the intestinal tract and respiratory system.

Because of the multiple origins of morphologically similar cells, and because a particular germinal layer may produce more than one type of epithelium, histological criteria are used to classify these cells.

Almost all epithelial surfaces are separated from the tissue they cover by a layer of acellular material. This layer, called the basement membrane, is a product of the connective tissue immediately below the epithelium. Unless special stains are used, the membrane is often difficult to see, but its presence is of importance to the histologist. The invasive potential of a malignant epithelium lies in deciding whether this membrane has been breached by the cells. Histological classification involves observation of the shape of the cells and their orientation with one another.

Shape

When height is compared with width, epithelial cells fall into three broad groups: squamous; cuboidal; and columnar.

Squamous cells—height negligible compared with width.

Cuboidal cells—height about the same as width.

Columnar cells—height is greater than width.

The types are more fully described later (see Normal Epithelial Cells, p. 37).

The free surface—the exposed part—of columnar cells may possess outgrowths of one or more fine hair-like projections, called cilia. Movement of these cilia may be controlled from within the cell, when they are said to be motile; when without this control they are termed non-motile.

Orientation

(a) *Simple epithelium*

An epithelial surface comprising a single layer of cells is called a simple epithelium (see Fig. 15); every cell rests on the basement membrane. All three types of epithelial cell can form simple epithelia; if a simple columnar epithelium carries cilia on its free surface, it is called simple columnar ciliated epithelium.

(b) *Stratified epithelium*

Epithelial surfaces that consist of more than one layer of cells are termed stratified epithelia (see Fig. 16); both squamous and columnar cells can show this form, the latter with or without cilia.

Squamous

Columnar

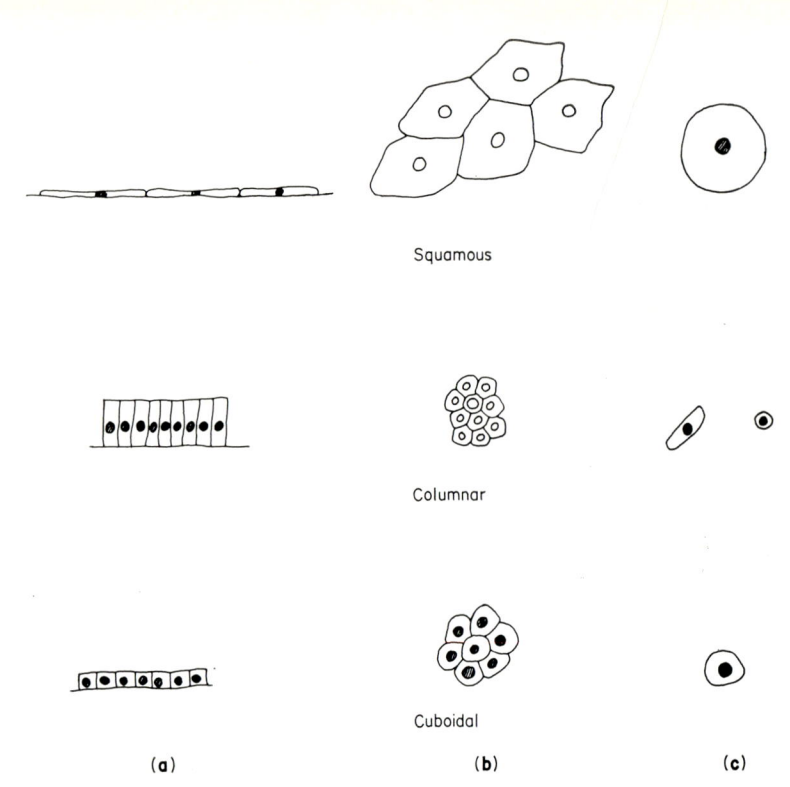

Cuboidal

(a)   (b)   (c)

FIG. 15. Diagrammatic structure of simple epithelia: (a), side view; (b), surface view; (c), shape of single cell in smear.

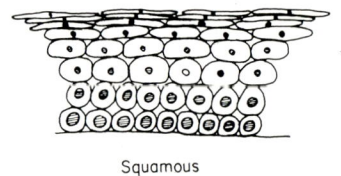

Squamous

Contracted   Stretched

Transitional - cuboidal

Stratified   Pseudostratified

Columnar

FIG. 16. Diagrammatic structure of stratified epithelia.

Transitional epithelium is a special type of stratified epithelium that lines the walls of hollow organs subject to stretching, for example, the urinary bladder. When fully stretched, only two layers of cells can be seen, a lower (basal) layer of cuboidal cells covered with a superficial layer of squamoidal cells: in contraction such tissue comprises many layers of irregularly shaped cells. This capacity to change morphology (pleomorphism) explains the wide variety of form the cells exhibit in smear preparations. These epithelia were called transitional because it was thought they represented an intermediate stage between columnar and squamous epithelia.

(c) *Pseudo-stratified columnar epithelium*

On section, this epithelium shows nuclei at different levels; the appearance is of tightly packed columnar cells, some of which are ciliate. Smaller triangular supporting cells are interspersed along the basement membrane with which all the cells have contact. Some histologists attribute a regenerative function to these small cells.

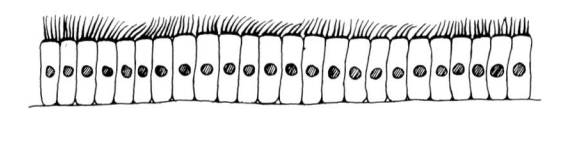

FIG. 17. Metachromal rhythm.

Stratification is usually seen in epithelia covering exposed surfaces where there is a continual loss of cells due to abrasion. Excessive loss of water and heat from the epithelial surface is prevented by the formation in the superficial cells of an impervious layer of keratin (a sclero-protein) before the nucleus loses its function. These cells are said to show keratinization. Squamous cells lining tissue in contact with isotonic and isothermal fluids, e.g., blood vessels, are not required to provide this protective function and therefore do not keratinize.

Tissues exposed to a stream of air, e.g., the respiratory tract, collect debris and are exposed to desiccation; these surfaces are lined with mucus-secreting and ciliated columnar epithelium. The mucus prevents desiccation and is propelled to the exterior by the action of the ciliate cells whose cilia move with a synchronized wave motion, termed metachromal rhythm (see Fig. 17). Mucus-secreting cells are also used to protect surfaces, e.g., the alimentary tract, by means of the lubricating and buffering properties of the mucus.

Normal Epithelial Cells

When the constituent cells of a tissue are dissociated from one another they quite often exhibit differences in morphology, compared with their

38　　　　　　　　　　　PRACTICAL CYTOLOGY

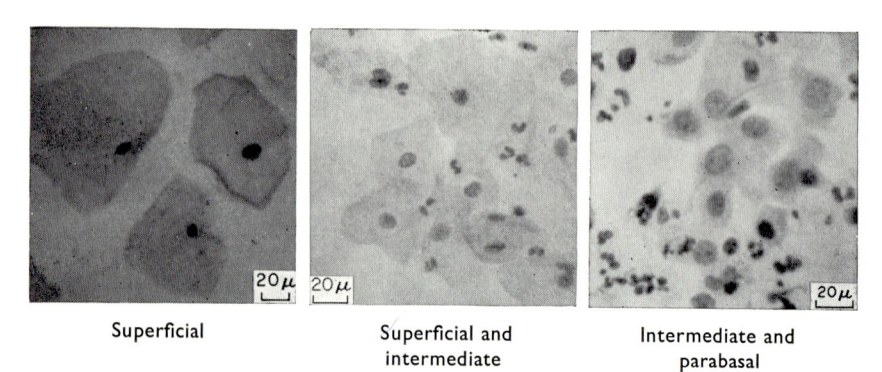

Superficial　　　　　Superficial and　　　　Intermediate and
　　　　　　　　　　　intermediate　　　　　　　parabasal

Stratified squamous cells

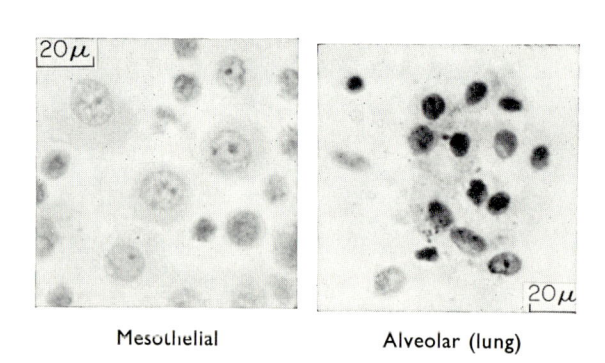

Mesothelial　　　　Alveolar (lung)

Simple squamous cells

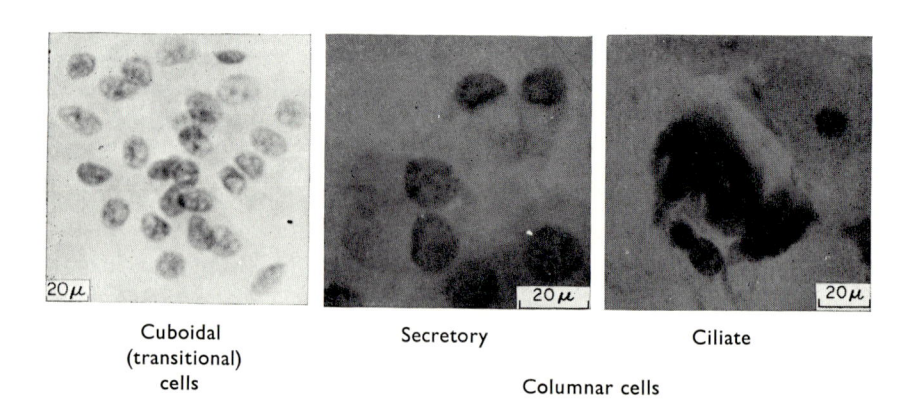

Cuboidal　　　　　Secretory　　　　　　　Ciliate
(transitional)
cells

Columnar cells

Fig. 18.

counterparts observed in tissue sections. In isotonic solutions, a single cell assumes that shape which results in equal pressure obtaining at all points on its limiting membrane. Cells with a uniform, non-rigid wall will become spherical, whereas in others, the non-rigid sections of the limiting membrane stretch until the tension is equilibrated with the pressure of the cell contents. Single cells frequently suffer further distortion during the preparation of the smear.

These two factors sometimes create difficulties in translating histological experience to "cytology smears".

Squamous Cells

(a) *Superficial cells*

The cells of some simple epithelia, e.g., mesothelium, lose their polyhedral shape when dissociated and appear in smears as round cells (see Figs. 18 and 19).

Cells of other simple squamous epithelia, e.g., those lining the lung alveoli, and the superficial cells of stratified squamous epithelium may resemble each other in their angulated, polyhedral shape.

Generally the height of the squamous cell is negligible in relation to its width. They are, therefore, seen in smears as flat, polyhedral cells with round, centrally placed nuclei. Some variation in size occurs depending on

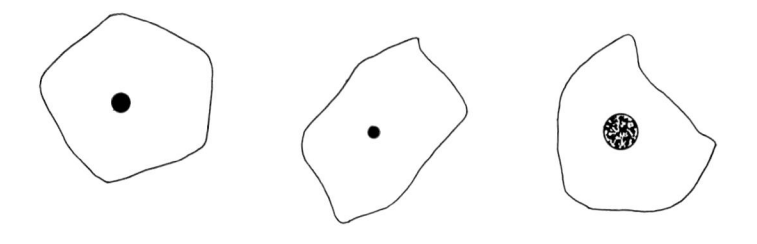

Superficial cells from stratified epithelium

○ Erythrocyte

Mesothelial cell

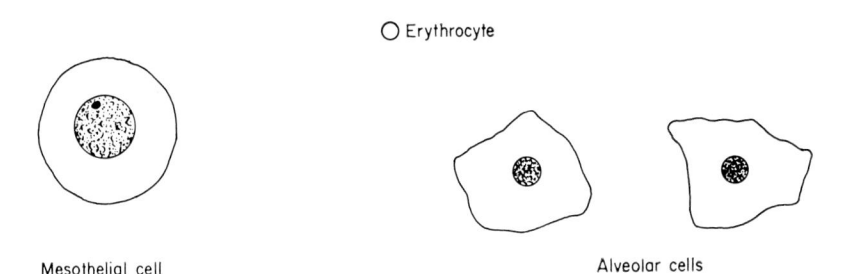

Alveolar cells

FIG. 19. Squamous epithelial cells seen in smears.

the site of origin; the smallest forms, however, are relatively large, about 30 μ across, and, when compared with other cells seen in smears, are easily identified. Plaques of these cells are frequently seen in material collected from a stratified squamous surface.

The nucleus varies in size between 8 and 14 μ, its chromatin exhibiting a fine, reticular pattern with a thin regular nuclear membrane. Cells arising from a stratified epithelium may possess nuclei whose chromatin has undergone pyknosis and karyorrhexis.

In a normal squamous cell the nucleus is completely contiguous with its adjacent cytoplasm. A feature of this cytoplasm is its homogeneous, almost transparent appearance. The limiting cell membrane is distinct and, in the superficial cell, well angulated. If the cytoplasm is folded, as sometimes happens in smears, the edges may present a serrated appearance. Cytoplasmic inclusions are not normally present. The cells lining the vaginal vault and buccal cavity may, however, contain glycogen granules, especially during pregnancy.

Smears collected from stratified squamous epithelial surfaces may contain squamous elements other than superficial cells; very occasionally the cells of the basal layer may be seen (see Figs. 20 and 21).

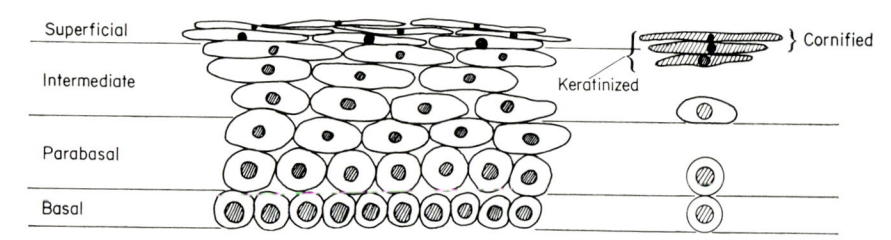

Fig. 20. Diagram showing structure and cells of stratified squamous epithelium.

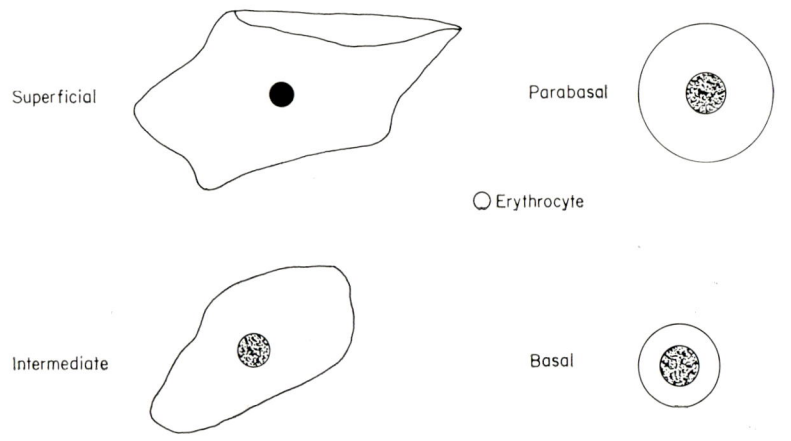

Fig. 21. Constituent cells of stratified squamous epithelium.

The successive layers of epithelial cells from the exterior toward the basement membrane are classified accordingly: superficial, intermediate, parabasal and basal cell layers. These divisions represent a transition from the replicating cell on the basement membrane, the basal cell, to the mature stratified cell.

(b) *Intermediate cells*

These cells lie immediately below the compact, superficial layer of squamous cells. Their immaturity is reflected in their smaller size and the characteristics of both cytoplasm and nucleus. Variation in cell size (anisocytosis) may be seen. The cytoplasmic border shows rudimentary angulation, but the cell is not keratinized and stains a uniform blue-green with Papanicolaou's solution. The round, regular nucleus (10 to 15 μ diameter) contains chromatin of even texture and the nuclear membrane appears as an even rim.

(c) *Parabasal cells*

These are round pale-staining cells, having a regular vesicular nucleus about 15 μ diameter. The cytoplasm shows no angulation and stains a uniform pale green. When non-angulated squamous cells are seen in smears they are usually of this class: the precursory basal cell is rarely seen.

(d) *Basal cells*

These are the germinal squamous cells and form a single layer resting on the basement membrane. In the smear they appear as relatively small (25 μ diameter) round cells, whose cytoplasm and nuclei stain much more intensely than the parabasal cell.

Variations occur in the staining of the cytoplasm in mature squamous cells; normal immature cells are basophilic (cyanophilic).

Keratinization and Cornification

In cytology these two terms are not synonymous. An epithelial cell whose cytoplasm is opaque and shows acidic (eosinophilic) staining properties is said to be keratinized. It may stain red, orange or yellow with Papanicolaou's stain.

A "cornified cell" refers to a squamous cell with a pyknotic nucleus (see Cornification Index, p. 76).

Cuboidal Cells

These are the cells constituting the transitional epithelium. The most frequent source of these cells—in a cytology laboratory—is the urinary system, especially the bladder.

Cells from these surfaces can exhibit a wide variation in size and shape, but the type most frequently seen is roughly ovoid with a centrally placed nucleus (see Fig. 22). Multi-nucleation (four to six nuclei) is common, and these nuclei may vary in shape, but they have a replicated chromatin pattern. Occasionally cells are seen having hyperchromatic nuclei, which are round and regular in shape and have chromatin of even texture.

The staining characteristics are imprecise; the nucleus varies in its chromatin content, and the cytoplasm ranges between strongly basophilic and strongly acidophilic.

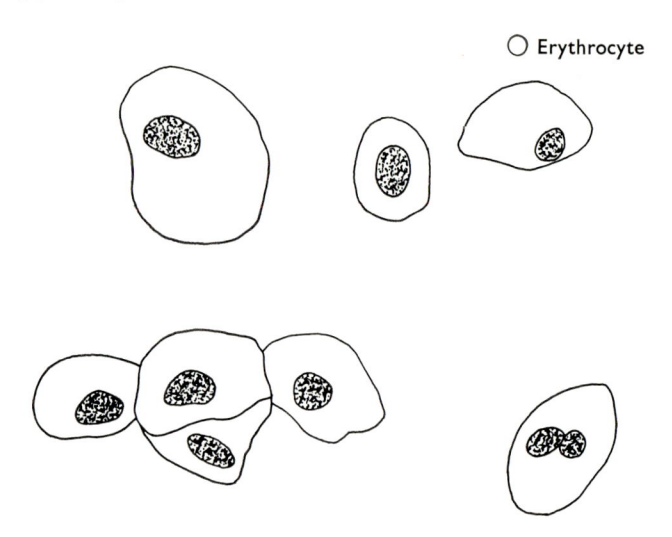

Fig. 22. Transitional - cuboidal cells.

Columnar Cells

(a) *General secretory type*

The significant difference between height and width usually causes single or small groups of these cells to be seen in side view in smears. Their size varies considerably according to site of origin. Some histologists have differentiated the typical, tall columnar secretory cell from much shorter forms present in some tissues, e.g., endometrium, describing the latter as cubical cells. In general terms, the columnar cell is a bullet-shaped cell, the pointed end constituting the base (see Fig. 23). Its nucleus, sub-central at rest, is round with chromatin of regular, open texture. The nuclear membrane is seen as an even thin rim. The position of the nucleus varies with the state of secretion. As the secretory granules increase in size the nucleus is depressed towards the base until, when the secretion is expelled, it resumes its sub-central position. In the resting cell the cytoplasm is seen as a very finely granular pale basophilic substance contained within a well

STRUCTURE OF THE CELL 43

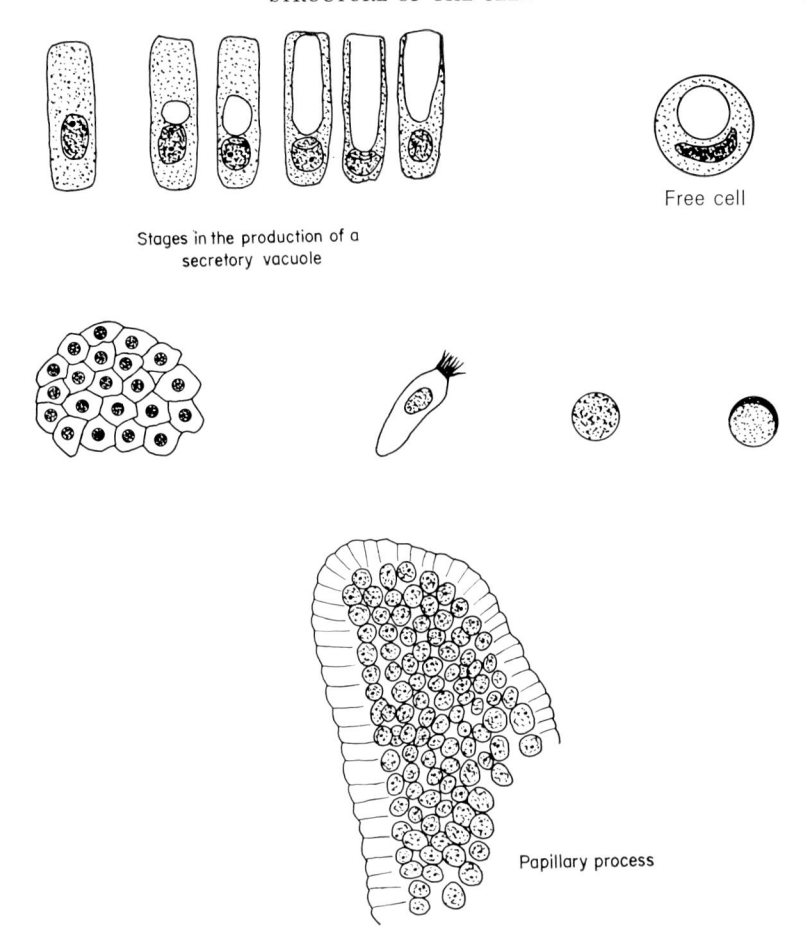

Stages in the production of a secretory vacuole

Free cell

Papillary process

FIG. 23. Columnar epithelial cells.

defined limiting membrane. Secretory vacuoles, if present, appear as unstained areas in routine preparations.

(b) *Ciliate type*

These cells are always present in association with secretory cells, sometimes interspersed between the latter as islands of cells. The cell approximates in shape to the secretory cell. Its exposed surface is flattened and noticeably thicker than the rest of the cell wall. The cilia are seen to arise from this flattened surface, the points of insertion being clearly defined. The number of cilia projecting from each cell varies with the origin of the cell; most carry from eight to ten. The nucleus is similar to that of the secretory cell.

Atypical Epithelial Cells

Epithelial cells deviating from normal morphology occur more frequently in smears as a result of a non-malignant (benign) process than a malignant (neoplastic) one. In cytology, these changes are sometimes referred to as benign and neoplastic atypia, respectively. The commonest, non-malignant, abnormal changes seen in the cells result from inflammation. Less commonly, changes may be due to benign aberrations in the tissue architecture, polyps and cysts for example, at or near the surface from which the smears have been taken. Certain diseases (e.g., vitamin B_{12} deficiency), therapeutic agents (e.g., radiotherapy) and ageing of the individual also induce changes that are characteristic of the stimulus and not of the location or type of epithelial cell. For most routine purposes, an ability correctly to interpret these benign atypia is related only to the necessity to differentiate them from neoplastic changes. Before discussing the effects of these stimuli on epithelial cells it will be helpful to examine the histological definitions of certain pathological changes that can occur in epithelial tissues.

Metaplasia

Metaplasia is a term used to describe the change in form and function of the cells of a tissue into a type not normally present, e.g., columnar into keratinized epithelium and fibrous tissue into bone. Transformation of columnar into stratified squamous epithelium does occur in the normal process of development of the oesophageal epithelium (Maximow and Bloom, 1948), but metaplasia is commonly associated with disease processes (malignant and benign). The change takes place in the regenerative and not the adult cells (Anderson, 1961). In cytology, we are primarily concerned with a metaplastic change in which the regenerative cells of simple columnar epithelium proliferate into a stratified squamoid type. This process is termed squamous metaplasia, and is usually associated with non-malignant lesions, e.g., chronic inflammation and polyps, although endometrial carcinomata not infrequently contain areas of squamous metaplasia. This type of metaplastic tissue is easily disrupted, and smears from such a surface may contain plaques as well as dissociated cells.

The cells can show various forms representing all stages in transition between the secretory columnar cells to a small, keratinized cell containing a nucleus that is either pyknotic or in karyorrhexis. Difficulty may be encountered initially in gynaecological smears in distinguishing between metaplastic cells and the abnormal squamous cells in certain "atrophic smears". Generally, the smaller size (about 20 to 25 μ diameter), replicate morphology and absence of bizarre karyorrhexis serve to characterize the metaplastic cells.

Neoplasia and Dysplasia

At a meeting of the International Committee for Histological Terminology convened in Vienna in August 1961, the following definitions were made:

(a) *Invasive carcinoma*

All lesions in which collections of epithelial cells infiltrate or destroy the underlying stroma are classified as invasive carcinomata. In this definition the extent of the invasion is irrelevant, but may be described in the report as minimal, extensive, etc. The microcarcinoma as a special subdivision has been deleted.

(b) *Carcinoma* in situ

The term "carcinoma *in situ*" applies to only those cases in which, in the absence of invasion, the total thickness of the epithelium appears de-differentiated. The same process can affect the lining epithelium of the cervical glands without being regarded as a separate category. It is admitted that the cells of the uppermost layers may show a slight degree of flattening. There are very rare cases in which a characteristic carcinoma *in situ* may show a greater degree of differentiation. These may be regarded as exceptions and cannot be classified separately.

(c) *Dysplasia*

All other disorders of differentiation of the squamous epithelium covering the surface and glands are to be classed as dysplasia. They may be described as "severe" or "slight" in preference to "doubtful" or "not doubtful", which are more subjective terms.

There is not complete agreement on the use of the above terms—some pathologists, for example, refer to carcinoma *in situ* as pre-invasive malignant epithelial dysplasia.

Dyskaryosis

The definitions above categorize histological deviation from the normal process of development of an epithelium although a rigid application of criteria is not always possible. A similar difficulty can occur in cytology when epithelial cells exhibiting abnormalities in development are termed dyskaryotic. These cells may be described as differentiated cells with nuclear and cytoplasmic abnormalities, notably increase in size.

Initially the observer is alerted by the presence of macrocytic, hyperchromic, atypical squamous cells, often in small plaques, which nevertheless match their normal counterparts in general morphology. Most striking are giant, hyperchromic superficial cells with well angulated cytoplasm. Although the nuclear - cytoplasmic ratio is within normal limits, anisonucleosis is apparent. The texture of the chromatin in many of the cells is a uniform opaque, blue or blue - black. Similarly affected intermediate and parabasal cells may also be seen.

Dyskaryotic cells arising from non-malignant lesions are not uncommon; for example, they are seen in monilia and trichomonas infections. Smears containing these cells demand careful searching and critical appraisal of the nuclear structure of affected cells in order to exclude malignancy.

Inflammation

Inflammatory agents usually induce increased exfoliation of cells from epithelia and also stimulate the replicating basal cells; the exfoliated cells show evidence of assault and abnormal numbers of immature cells are seen. In general terms the effects of inflammatory agents are:

1. To stimulate regenerative basal cells, sometimes producing cells with abnormal characteristics, e.g., multinucleation.
2. To cause swelling of epithelial cells with functional nuclei.
3. To disturb the mechanism of chromatin condensation in epithelial cells which normally pyknose.
4. To produce a stimulus that attracts non-epithelial cells to the focus of the assault.

"Inflammatory" cells showing multi-nucleation usually possess from two to five nuclei that have a normal chromatin pattern although the constituent nuclei may vary in size.

Swelling of the cells appears to be due to imbibition of water and the increase in size results in a reduction in staining intensity. This uptake of water may cause the cell wall to rupture, thus allowing the nucleus to float free ("naked nuclei"). Distension of the nuclear membrane is associated with precipitation or sometimes laking of the chromatin. Particular inflammatory agents, especially certain infections, often produce characteristic chromatin patterns. Superficial squamous cells usually remain intact, but the nucleus may be either swollen or shrunken and wrinkled with altered chromatin.

The more severe the damage to an epithelial surface the more difficult becomes the process of repair. When the replicating basal cells are destroyed, exposing the underlying connective and vascular tissues, an ulcer or an erosion is produced and is seen as a raw area that may exude blood and pus. Erosions may become covered during healing with metaplastic tissue.

Polyps

Columnar epithelium may undergo metaplastic change in conditions other than inflammation, e.g., in the formation of polyps. These are tumours, usually benign, consisting of a central core of mesodermal tissue (stroma and blood vessels) surrounded by the proliferating epithelium. The surface is usually covered with a layer of metaplastic cells that is easily damaged: when this occurs, the exposed surface is quickly colonized by bacteria. Mucin-secreting cells within the polyp usually cause the formation of small cysts. Smears taken from these lesions may contain plaques of columnar cells that exhibit a regular "honeycomb" structure. Most plaques are of tall columnar cells, but others showing metaplastic change may be seen. Multi-nucleated secretory cells are frequently seen, usually as large pale-staining cells containing two or three replicate nuclei. Blood is often present, together with numerous neutrophils and histocytes.

Other Non-malignant Causes

Squamous epithelial cells and the columnar cells of the gastric mucosa exhibit abnormal changes in vitamin B_{12} deficiency. Increase in size of the whole cell occurs, but the ratio of nucleus to cytoplasm remains within normal limits. Occasionally multi-nucleation and irregular chromatin patterns are present (Graham, 1963).

The effects of ageing and of radiotherapy are frequently encountered in gynaecological smears, and a description of these changes is given in the Section on gynaecological smears (ageing, p. 74; radiotherapy, p. 84).

Malignancy

As a result of work by Papanicolaou and others, visual tests have been formulated to differentiate malignant from non-malignant cells. The following points are criteria that can be applied to most malignant cells seen in smears:

1. The ratio of nucleus to cytoplasm is altered.
2. When the nuclear chromatin is not condensed, it shows a "blotched" pattern, often with irregular cleared areas.
3. There is variation in nuclear size (anisonucleosis) and staining intensity of chromatin (anisochromia) among the abnormal cells.
4. The nuclear membrane is irregular and shows abrupt variation in thickness.
5. The nuclear chromatin shows sharp points and angles.
6. The nuclei show abnormal nucleoli—more than the normal number or increase in size.

The first criterion is properly placed: it pre-supposes a clear knowledge of the morphology of all the normal cells likely to be encountered. This can only be gained by meticulous examination of many slides and constant reference to authoritative opinion. The remaining criteria reflect the level of our understanding of the malignant nucleus and their variety serves to emphasize the hazard of reporting on conclusions drawn from a few abnormal cells.

Some cytologists (Kasdon and Bamford, 1962) have defined precise values for the nucleus - cytoplasm ratio in different epithelial disorders, but application of these can be fallacious. As experience is gained, many visual criteria are applied, some perhaps unconsciously, and it is the computation of the results, after the criteria have been applied to all the abnormal cells seen in a smear, that directs the worker's opinion.

Classification of Morphological Changes: Nuclear Changes

In order to examine the atypical changes occurring in epithelial cell nuclei we can divide the cells broadly into two groups:

1. Cells with open, functional nuclei.
2. Cells whose nuclei are condensing or about to condense.

The changes described below are illustrated in Fig. 24 and summarized in Fig. 25.

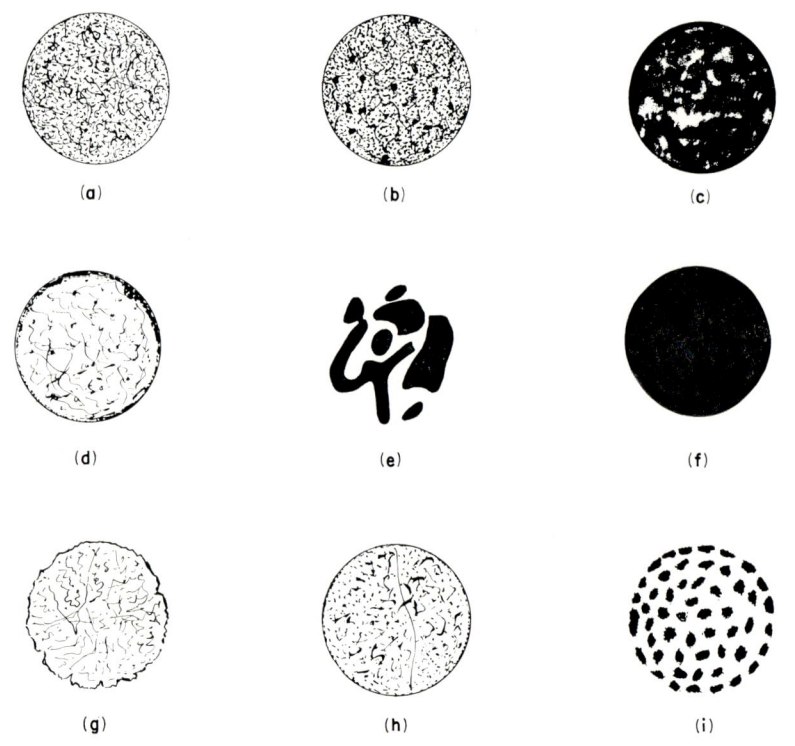

FIG. 24. Benign atypical changes in nuclear chromatin of epithelial cells; (a), normal; (b), hypochromia coarsening; (c), skeining; (d), hypochromia central pallor; (e), karyorrhexis; (f), hyperchromia; (g), crenation; (h), folding; (i), karyolysis.

STRUCTURE OF THE CELL 49

FIG. 25. **General guide to some distinguishing cell features:** − **signifies usually absent;** ± **signifies sometimes present;** + **signifies usually present.**

Group 1. Functional Nuclei

(a) *Nuclear membrane*

The thickness of the nuclear membrane is normally uniform, so that when the cell is viewed from above the nucleus is seen as a round mass limited by a regular rim of apparently darker-staining material. The dark ring is due to the greater depth of nuclear membrane that is being viewed at the circumference. Epithelial cell nuclei showing inflammatory change usually exhibit a regular nuclear membrane, whereas gross variation in thickness occurs in malignant nuclei.

(b) *Chromatin texture*

When fixed, the matrix of nucleoprotein of the normal, resting epithelial cell appears as a finely recticular, blue-staining mass, except for the spherical nucleolus, if present, limited by the nuclear membrane. Inflammatory agents frequently cause the chromatin to precipitate in a floccular or granular (fine or coarse) form as regular or irregular sized particles: occasionally laking of the chromatin occurs. Usually numerous epithelial cells in the smear are affected, and there is a general uniformity in the pattern of the abnormal change. The irregular chromatin precipitation seen in malignant nuclei varies considerably from cell to cell and, frequently, large irregular clear areas are seen.

(c) *Hypochromia*

Nuclei staining less intensely than their normal counterparts are said to show hypochromia.

Benign hypochromic nuclei show a marked reduction in staining intensity of a regular, but coarser than normal, chromatin pattern. A thick even nuclear rim is distinctly seen. The cell, frequently larger than normal, often shows an overall hypochromia of both nucleus and cytoplasm. A characteristic central pallor is sometimes seen in all the nuclei of cells in smears of the vaginal exudate in trichomoniasis. Smears from neoplastic lesions often contain examples of cells with markedly hypochromic nuclei. This is probably a result not so much of an increase in nuclear size but because of a reduced number of chromosomes in the nucleus.

(d) *Hyperchromia*

Nuclei are said to exhibit hyperchromia when they possess an increase in chromatin, compared with similar normal cells. Cells, both larger and smaller than their normal counterparts, may also be hyperchromic in respect of both nucleus and cytoplasm. Squamous cells showing these features, in association with increased density of cytoplasm, contribute to the dyskaryotic cytology arising from a dysplastic epithelium. Malignant cells frequently show unevenness in texture of the hyperchromic nucleus, which often has a hard, glassy surface, in contrast to the matt appearance of benign nuclei.

(e) *Anisonucleosis*

This term is used to describe variation in nuclear size in cells of the same type. The most frequently seen example occurs in columnar cells. These cells respond quickly to assault and there is an increase in size of the whole cell, including the nucleus, with a consequent reduction in staining intensity. Anisonucleosis is also a common characteristic in transitional cells. In contrast to these nuclei, which have an even chromatin texture, anisonucleosis in malignancy is associated with gross irregularities in the chromatin pattern.

(f) *Naked nuclei*

If the cell osmoregulatory mechanism breaks down, its limiting membrane may rupture and the nucleus will float free, sometimes with adherent fragments of the cytoplasm. This commonly occurs in columnar cells, but is also frequently seen in immature squamous (parabasal) cells. The nuclei may be present singly, in small clusters, or large groups, and can usually be identified as originating from a particular type of epithelial cell. In many instances the chromatin pattern is lost. Laking of the chromatin may occur, giving the naked nucleus a glazed, blue appearance, if the cells are of columnar type, whereas the nuclei derived from so called "atrophic" (parabasal) squamous cells generally stain with a translucent or opaque matt grey colour. In malignant cells which appear to possess no cytoplasm, the chromatin usually stains avidly, often with a grossly irregular pattern.

(g) *Multi-nucleation*

Normally an epithelial cell contains a single nucleus. The production of cells containing two or more replicate nuclei is a common response in inflammation. Columnar and transitional epithelia often react in this way, producing cells with two or more nuclei having a normal chromatin pattern, which may vary in size but not in chromatin texture. In inflammatory smears, mature squamous cells may be seen with two nuclei, and mesothelial cells may contain up to five. Cells from neoplastic lesions, polyps and cysts may contain multiple nuclei exhibiting considerable variation in size and chromatin pattern.

(h) *Nucleoli*

These structures, which are often observed in mesothelial and secretory (columnar) cells, are not commonly seen in the nuclei of squamous and transitional cells. When present, they appear as single, reddish-blue or dark-blue spheres, about 3 μ in diameter, placed eccentrically within the nucleus. The presence of large or multiple nucleoli is an atypical feature. In malignancy, nucleoli exhibit a considerable variation in size between cells.

Group 2. Pyknosing Nuclei

Cells with nuclei that are undergoing or about to undergo condensation may exhibit additional abnormal changes.

(a) *Pyknosis*

When a nucleus stains as a structureless mass of chromatin, dark blue or black with haematoxylin, it is said to be pyknotic. This process, which occurs normally in some stratified squamous and transitional cells, may also be seen in columnar cells that have undergone metaplasia. Smears containing malignant cells usually show at least one or two examples of pyknotic neoplastic cells, which may be microcytic or macrocytic when compared with the normal.

(b) *Karyorrhexis*

This stage of chromatin condensation precedes the disappearance of the nucleus. An epithelial cell that does not normally experience karyorrhexis may do so under abnormal circumstances. When this happens, the fragmenting chromatin is seen as quite irregularly sized particles. Examples of this may be seen in the vaginal squamous cells in atrophic vaginitis.

(c) *Karyolysis*

This abnormal disintegration of condensed chromatin occurs commonly in respiratory-tract infection and less frequently, in cervicitis. The chromatin coalesces into tiny regular spherules, evenly spaced so as to indicate the original shape of the nucleus: the nuclear membrane disappears.

(d) *Folding and crenation*

An open nucleus may possess a single fold, or several irregular wrinkles, on what would normally be a spherical limiting membrane. Such a nucleus is said to be crenated. Although the chromatin retains its open pattern, its texture is noticeably coarser; this appears to be due to precipitation of the reticulum of nucleoprotein as fine, regular granules. The nuclear membrane is usually thin, distinct and of even thickness. Squamous cells whose nuclei exhibit this characteristic are commonly seen in "inflammatory" smears. It is not a usual feature of malignant cells, although convolutions in the nuclear membrane, described as "craggy" (Philps, 1964), may occur in neoplastic squamous cells in lung, and the precipitation of chromatin in ridges, described as "trabeculation" (Kasdon and Bamford, 1962), may occur in neoplastic squamous cells originating in the cervix uteri.

Classification of Morphological Changes: Atypical Cytoplasmic Changes

Orientation of Nucleus

A non-malignant atypical cell generally retains its nucleus in the same position as its normal counterpart; this is central in most cases. In malignancy, numerous examples can be seen of cells whose nuclei are disorientated.

Nuclear Halo

Normally the nucleus and its adjacent cytoplasm are contiguous. Cells that show a clear, unstained zone around part, or all, of the circumference of the nucleus are, respectively, described as possessing a partial or perinuclear halo. Squamous cells, intermediate and superficial, from inflamed sites often exhibit this halo, which is not a feature of other benign or malignant cells.

Hyperchromia

The cytoplasm of many atypical epithelial cells is hyperchromic and less transparent than normal.

Cell Wall

Normal epithelial cells have a clearly seen boundary to the cytoplasm, although the cell wall itself may not be visible. Abnormal cells, both malignant and benign, may produce a thin or irregular limiting membrane and cause the cell boundary to appear indistinct. At times exfoliated squamous cells, otherwise normal, may show a ragged, indistinct border that is usually the result of bacterial action; some nuclei may have only a residual tag of cytoplasm. This appearance is described in gynaecological cytology as a "cytolytic picture".

Inclusions

Inflammatory lesions frequently produce epithelial cells with intracellular neutrophils. These inclusions are surrounded by a thin-walled sac containing an unstained, transparent fluid. As many as twenty or more may be seen in a single squamous cell. The feature may be seen in "malignant" smears, although the number of neutrophils is smaller (one or two per cell). Occasionally one neoplastic cell may appear to have engulfed another (bird's eye cell).

Adherent Neutrophils

Superficial squamous cells are quite often seen in purulent gynaecological smears with many neutrophils adherent to the cytoplasm; other squamous cells in the field are unaffected. Close scrutiny of these other cells will reveal "pitting" of the cell wall in a few of them. This picture is associated with trichomoniasis.

Vacuolation

Single vacuoles in the cytoplasm of cells are commonly seen; usually they are manifestations of an active cell (see The General Cell, p. 26). Cells with multiple vacuoles in both cytoplasm and nucleus are generally considered to be degenerate.

Many of the nuclear and cytoplasmic changes described above are illustrated in Figs. 26 and 27.

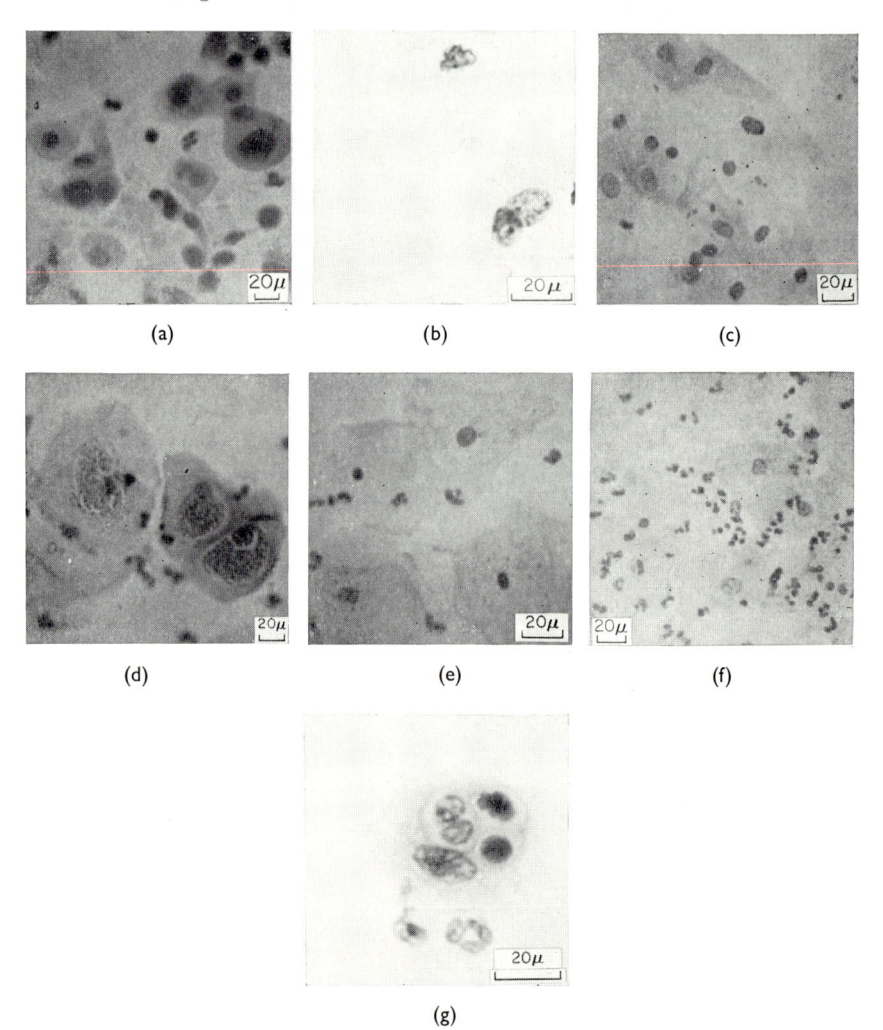

FIG. 26. (a), atypical cells from polyp; (b), squamous cell showing coarsening of chromatin (inflammatory change); (c), squamous cells showing nuclear haloes; (d), squamous cells containing ?glycogen inclusion; (e), dyskaryosis (monila infection); (f), dyskaryosis (trichomonas infection); (g), squamous cell, showing neutrophil inclusions.

(a)   (b)

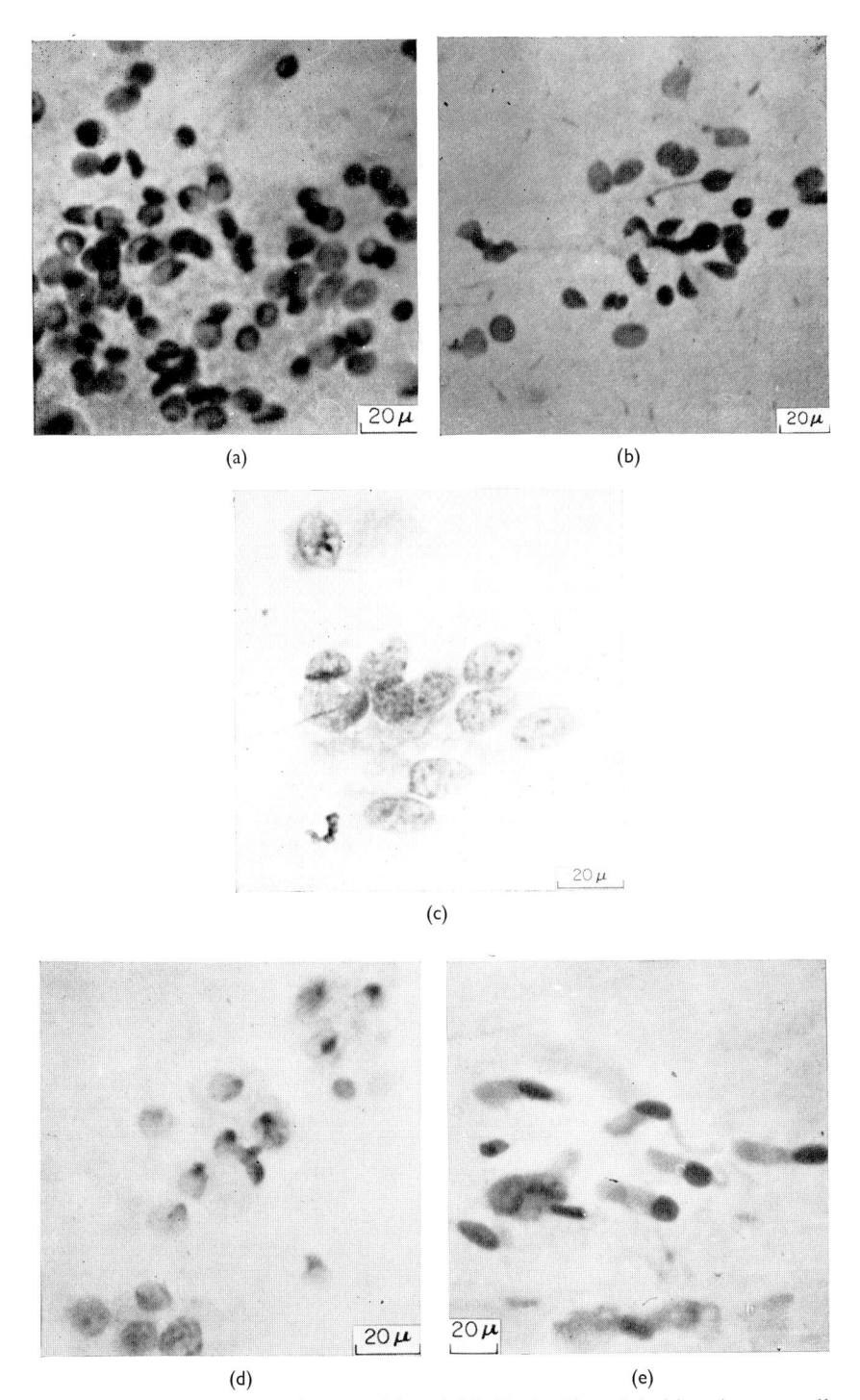

(c)

(d)   (e)

FIG. 27. Benign nuclear changes: (a) and (b), "naked" nuclei; (a), columnar cells with laked chromatin; (b), atrophic squamous cells; (c), columnar cells with swollen nuclei; (d), disintegrating columnar cells; (e), columnar cells showing nuclear pyknosis.

Designated Cells

The cells described below are illustrated in Fig. 28.

Mitotic Figure

The nuclei of cells in mitosis are sometimes referred to as "mitotic figure". Generally speaking, it is not possible to classify mitosing cells, but their presence need not cause difficulty. It is not common to find mitosing cells in "malignant smears"; they are, however, often seen in smears containing evidence of increased histocytic activity.

Fibre Cell

This name is given to long strap-shaped cells with sausage-shaped nuclei. Malignant cells of this type are sometimes seen in smears from a cervical squamous carcinoma. Graham (1963) regards their presence as a strong indication that the lesion is invasive. Morphologically similar cells may be found in smears from a dysplastic (non-malignant) epithelium and, unless the chromatin conforms to the criteria for malignancy (see Malignancy, p. 47), judgment should be reserved.

Signet-ring Cell

A cell that possesses a single, cytoplasmic vacuole, sufficiently large to distort the nucleus and cause its displacement to the periphery of the cell, is described as a signet-ring cell. The term has been applied to affected parabasal cells (Smolka and Soost, 1965); histiocytes (Koss and Durfee, 1961) and gastric epithelial cells (Graham, 1963). Cells exhibiting this signet-ring morphology are also seen in smears of pleural and ascitic fluids.

Tadpole Cell

Occasionally, a parabasal cell may be seen whose cytoplasm has been extended at one point forming a long tail. Although cells with this shape are atypical, they are not necessarily malignant. Classification must rest on irregularities in the chromatin (see Malignancy, p. 47).

Epithelial Pearls

Whorled clusters of cells, wrapped round one another like the layers of an onion, may occasionally be seen in smears from stratified squamous epithelium. These structures are called "epithelial pearls" and may originate in either benign or neoplastic epithelia. The size of the cluster may vary between a few (8 to 12) and several (up to 40 or so) cells. If the constituent nuclei cannot be firmly identified as malignant, and no other malignant cells are seen, their presence need not arouse suspicion.

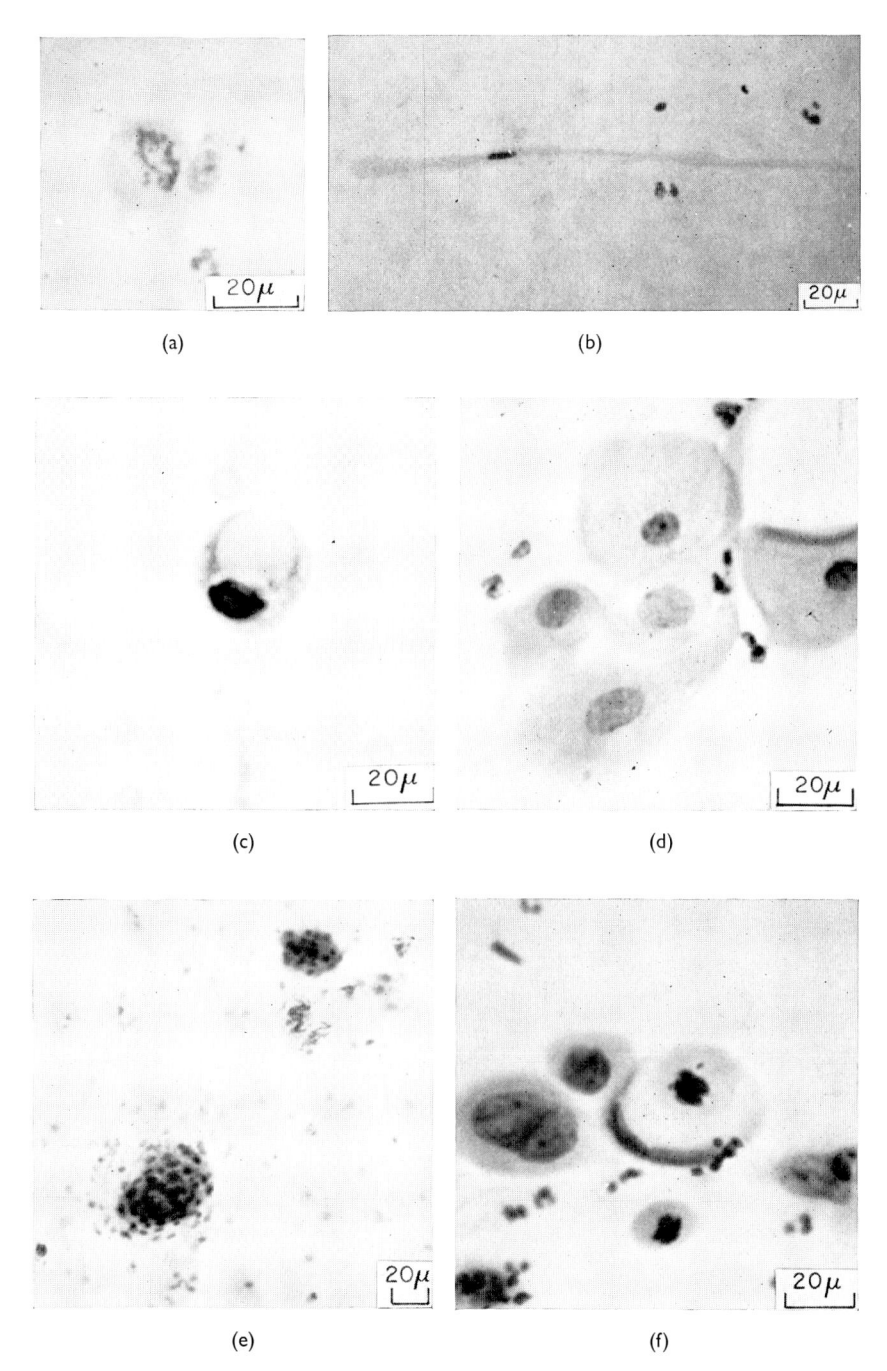

FIG. 28. Designated cells: (a), mitotic figure; (b), fibre cell; (c), signet-ring cell; (d), tadpole cell; (e), epithelial pearl; (f), bird's eye cell.

Bird's Eye Cell

This description is given to a malignant squamous cell that possesses a second squamous cell as a cytoplasmic inclusion. The size of the included cell causes displacement and distortion of the host nucleus. Although the cause of the phenomenon is not known, a commonly offered explanation attributes a "cannibalistic" propensity to the enclosing cell.

Papanicolaou Grading

Having established the presence of abnormal cells in a smear, the conjecture that they may have originated from a neoplastic focus naturally arises. A technician refers such smears to the cytologist, who will then make his report. A generally accepted routine is to follow up a suspicious or positive cytology report with surgical biopsy of diseased tissue. The subsequent histological examination will explain the cytological findings and, if this lesion is neoplastic, whether it is invasive or pre-invasive.

A very short experience examining smears is sufficient to show the worker that a decision concerning cells is not always easy. When several preparations have been carefully scrutinized in association with a tutorial of catalogued cells, it is possible to classify a smear as "normal". Unfortunately, the number of such smears found in a random selection of patients will be very small; in gynaecological cytology, for instance, a good deal less than 1% will fall into this group. So that although classifying smears into the two categories, normal and "abnormal", is relatively easy, it does not achieve a great deal. Further experience, again in association with a tutorial, will allow the worker to detect smears containing cells that are obviously malignant. The number of this type of smear will also be small, leaving a large intermediate group. Most of these will contain atypical cells that, with experience, can be attributed to various agencies other than malignancy. There still remains a group, however, whose classification will cause disagreement even between experienced workers. Largely because of this difficult group, a method of reporting smears was devised by Papanicolaou, who used a numbered grading system to indicate his opinion. The meaning of the five grades is as follows:

1. Absence of atypical or abnormal cells.
2. Atypical cytology, but no evidence of malignancy.
3. Cytology suggestive of, but not conclusive of, malignancy.
4. Cytology strongly suggestive of malignancy.
5. Cytology conclusive for malignancy.

Subsequent workers in cytology have modified the grading in various ways. Perhaps the most important to notice is the introduction of an "0" grade to indicate an unsatisfactory smear.

Row A

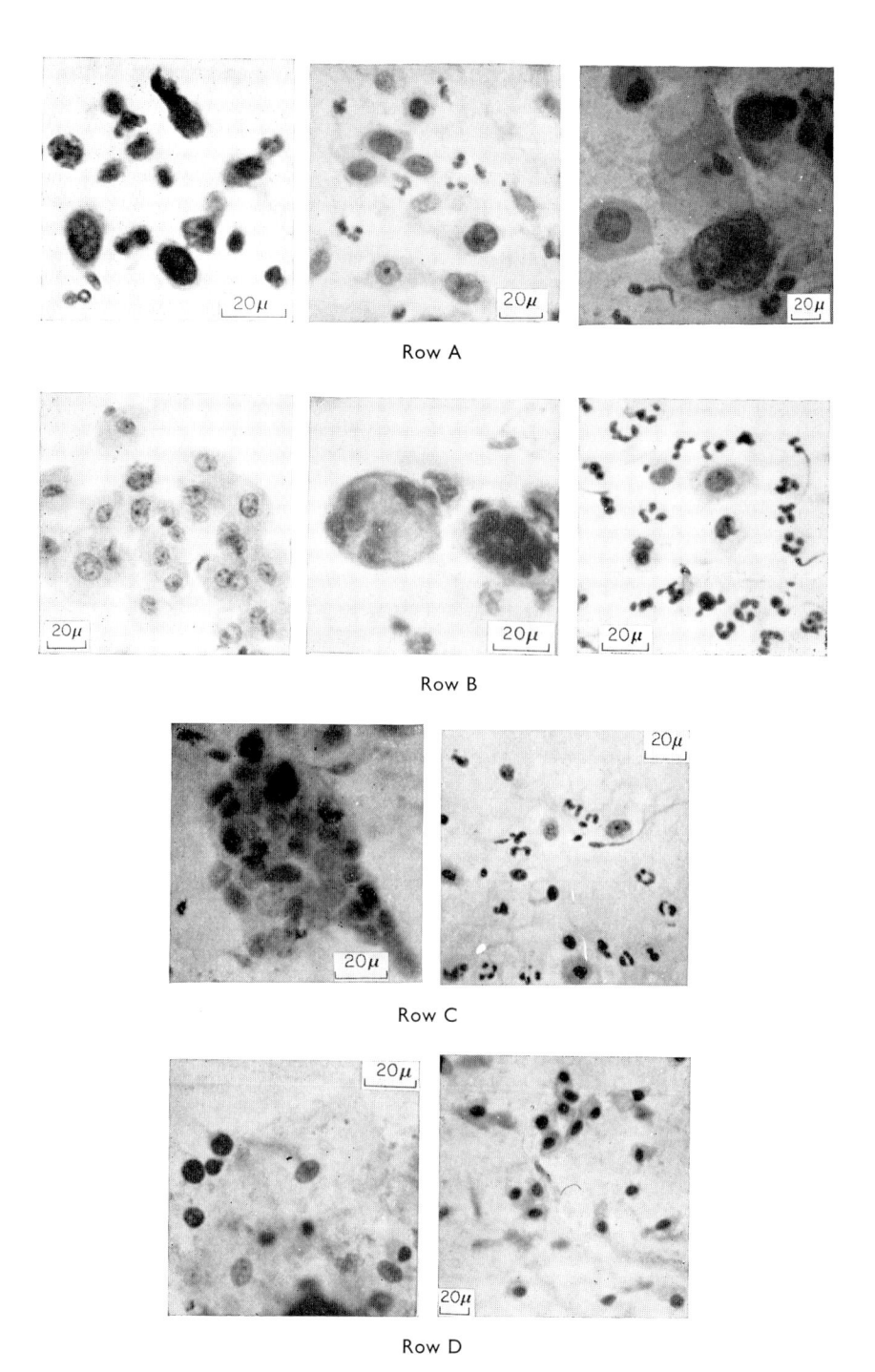

Row B

Row C

Row D

Fig. 29. Rows A and B, malignant squamous cells: row A, gynaecological smears; row B, sputum smears. Rows C and D, malignant columnar cells: row C, gynaecological smears; row D, sputum smears.

Except in circumstances so infrequent as to be ignored, a decision to report a smear as "positive" will not be based on the finding of a single or a few atypical cells. If, after careful examination, the smears contain only one or two apparently malignant cells, the logical step is to request further smears. A carcinomatous area exfoliating atypical cells is unlikely to produce repeat smears yielding a second equivocal result, provided the area has been scraped. It is a feature of an *in situ* carcinoma of the uterine cervix, for example, that atypical cells can be numerous in the smears even when subsequent serial histological sections show only a small area of diseased epithelium. Some examples of malignant cells are given in Fig. 29.

Cytology can be difficult in some clinically obvious neoplasms. Smears from such lesions may contain a few necrotic (and, therefore, poorly stained) cells in a matrix of blood, and the paradoxical circumstance occurs when the smears cannot confirm the clinical finding. The frequency of this anomolous type of smear is not great.

Most "difficult" smears fall into two categories. The first, and most frequent, group contains atypical cells, most of which are classifiable, but a proportion of which are not. The few have "overstepped the mark" and closely approach all the criteria of a malignant cell. The second group contains cells that are virtually unclassifiable because they exhibit such a wide range in morphology and other structural characteristics. Occasional cells in these smears may seem almost to conform to the criteria for malignancy, but some doubt remains.

Examples in gynaecological smears of the two categories are, in the first instance, some Trichomonas infections and, in the second, smears containing cells from a cyst or polyp and also smears from an infected atrophic epithelium. Repeating such smears, after treatment where necessary, often resolves the problem.

Reporting slides by numbered grading is convenient, provided the person receiving the result is completely conversant with the grades. It is often arranged that on receipt of a report of a slide with a grading above 2, a repeat specimen is automatically supplied.

Cataloguing slides by using the Papanicolaou grading is a necessary hazard to the initiate in cytology. A worker without ready access to experienced opinion must be continually on his guard against using the 3 grading as a refuge.

Studies of neoplastic cells examined *in vitro* (Boddington, *et al.*, 1965; Wakonig-Varrtaja and Hughes, 1965) may yield more precise methods for identification, but until these have been designed and tested cytologically, reporting will continue to depend on experienced opinion.

Slides from a series of 1,500 apparently well women were examined and an analysis of results is shown in Table I. Subsequent histological investigation of patients whose smears were graded 4 and above showed:

One false positive, the cells having arisen from an infected epithelium.
Ten squamous carcinomata (nine pre-invasive and one invasive).
One adenocarcinoma.

An additional histological diagnosis of adenocarcinoma has been made on a further patient in the series whose smear grading in the "well women" clinic was 2. This underlines the necessity for careful scrutiny of smears collected at these centres. The difficulty of recognizing adenocarcinoma in gynaecological cytology smears is emphasized when the figures for uterine cancer diagnosed histologically at St Luke's Hospital, Bradford, during 1964 and 1965 are analysed (McInroy, 1966). This analysis shows the ratio of squamous carcinoma (invasive and pre-invasive) adenocarcinoma to be 2 to 1, and the accuracy of cytological diagnosis, during the period reviewed, to be 97% for squamous carcinoma and 67% for adenocarcinoma.

Examination of the results in Table I shows that more than 90% of reports involve the 2 grading. Most of these smears contained epithelial cells affected by inflammatory changes. The high incidence of this "inflammatory smear" shows that the "normal grade 1 smear" is the exception rather than the rule and has led latterly to the linking of grades 1 and 2 in reporting these "normal" smears.

TABLE 1

Grading analysis of 1,511 apparently well women

(Bradford Hospital, 1965-1966)

Papanicolaou Grading	Number of Cases
0/1	7
1	108
1/2	268
2	1,105
2/3	0
3	0
3/4	1 ⎫ 4
4	3 ⎭
4/5	7
5	1

Non-epithelial Cells Seen in Smears

Most of these cells can be studied by preparing smears of buffy-coats from several whole-blood specimens. The smears should be prepared by using alcohol - ether fixation and the haematoxylin - Papanicolaou staining sequence.

All the cells are of mesodermal origin arising from the tissues of the reticulo - endothelial system (including the lymphopoetic and the haemopoetic systems).

The reticulo - endothelial system produces:
  Histiocytes and the monocytes of blood.
  Plasma cells.
  Fibroblasts.
The lymphopoetic system produces:
  Lymphocytes.
The haemopoetic system produces:
  Granulocytes (neutrophils, eosinophils and basophils).
  Erythrocytes.

With the exception of histiocytes and neutrophils, these cells are not normally present in an epithelium or on its surface.

Histiocytes (Macrophages)

These amoeboid cells can move freely through the body tissues in response to chemotaxis, reproducing themselves by mitosis if called upon so to do. Their function is to remove substances foreign to the tissue by surrounding and engulfing the material (phaging), which ranges from complex, organic matter, for example dead whole cells, to inorganic substances, such as talc or carbon. This activity is reflected in its pleomorphism. The diversity of appearance can cause difficulty on occasions; its size may vary between 8 and 25 μ diameter, and sometimes greater. Some cytologists have adopted an arbitrary classification of the histiocyte into three morphological types: small, large and multi-nucleate.

General structure

The nucleus of the cell has a characteristic indentation that often confers on it a kidney (reniform) shape (see Figs. 30 and 31). This indentation, referred to as a "hof", can generally be identified by focusing through the cell. Irregular aggregates of chromatin are present in the nucleus, and, in active cells, large cleared areas may be seen. Although not always sharply defined, the nuclear rim is seen to be of even thickness. Generally, the nucleus is eccentrically placed within a cytoplasm that exhibits wide differences in staining and texture. It may appear uniformly translucent, finely fenestrated or containing one or more large unstained vacuoles. The affinity of the cytoplasm for the Papanicolaou stains is not great and it may show weak cyanophilic or eosinophilic properties. A characteristic common to histiocytes is the indistinct limiting membrane of the cell.

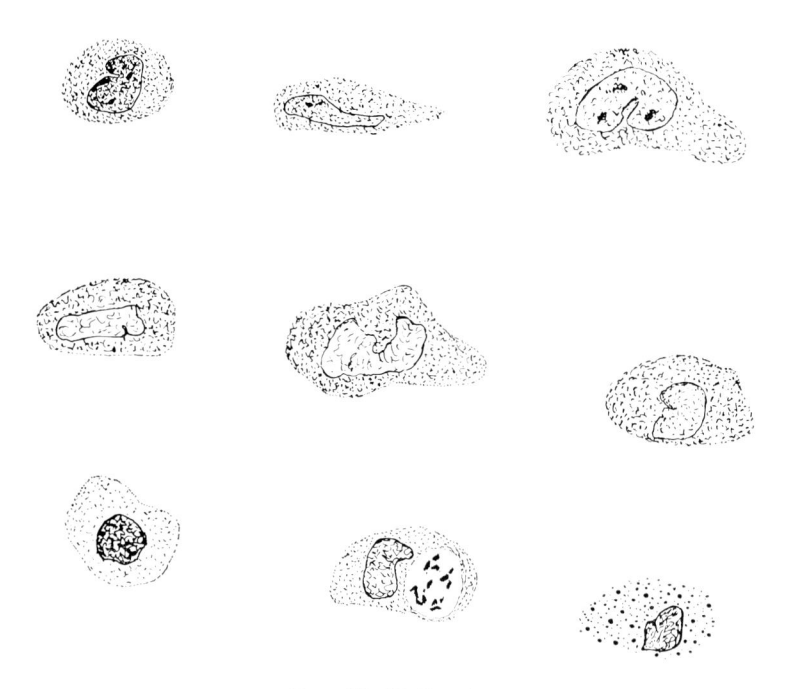

FIG. 30. Histiocyte.

The phagocytic properties of the histiocyte frequently helps in identification. An entire squamous epithelial cell may be ingested, which, of course, may be considerably larger than the resting histiocyte itself. Histiocytes may also be seen whose cytoplasm contains one or several neutrophils. Cells engulfed by the histiocyte often show degenerative changes, the result of digestion. A noticeable feature is the striking decrease in size of the ingested cell, which, despite this, retains, for some time, its other general morphological characteristics. Other material ingested by these cells include starch granules, inorganic powders (e.g., from talc) and, in the respiratory tract, carbon particles from inhaled soot. Histiocytes frequently die as a consequence of phagocytosis. The cytoplasm fragments, although small tags, may remain attached to the nucleus. In these circumstances, the chromatin may lake and stain a uniform lavendar colour with the Papanicolaou stain.

When large numbers of histiocytes are present on an epithelial surface, evidence of this increased activity is seen in the smear. Juvenile histiocytes are seen clumped together, with an occasional cell in mitosis, and others may be seen containing multiple, replicate nuclei. The latter are sometimes described as "giant histiocytes", "pseudo-giant cells" (Smolka and Soost, 1965) or "foreign-body giant cells" (Papanicolaou, 1954).

(a)

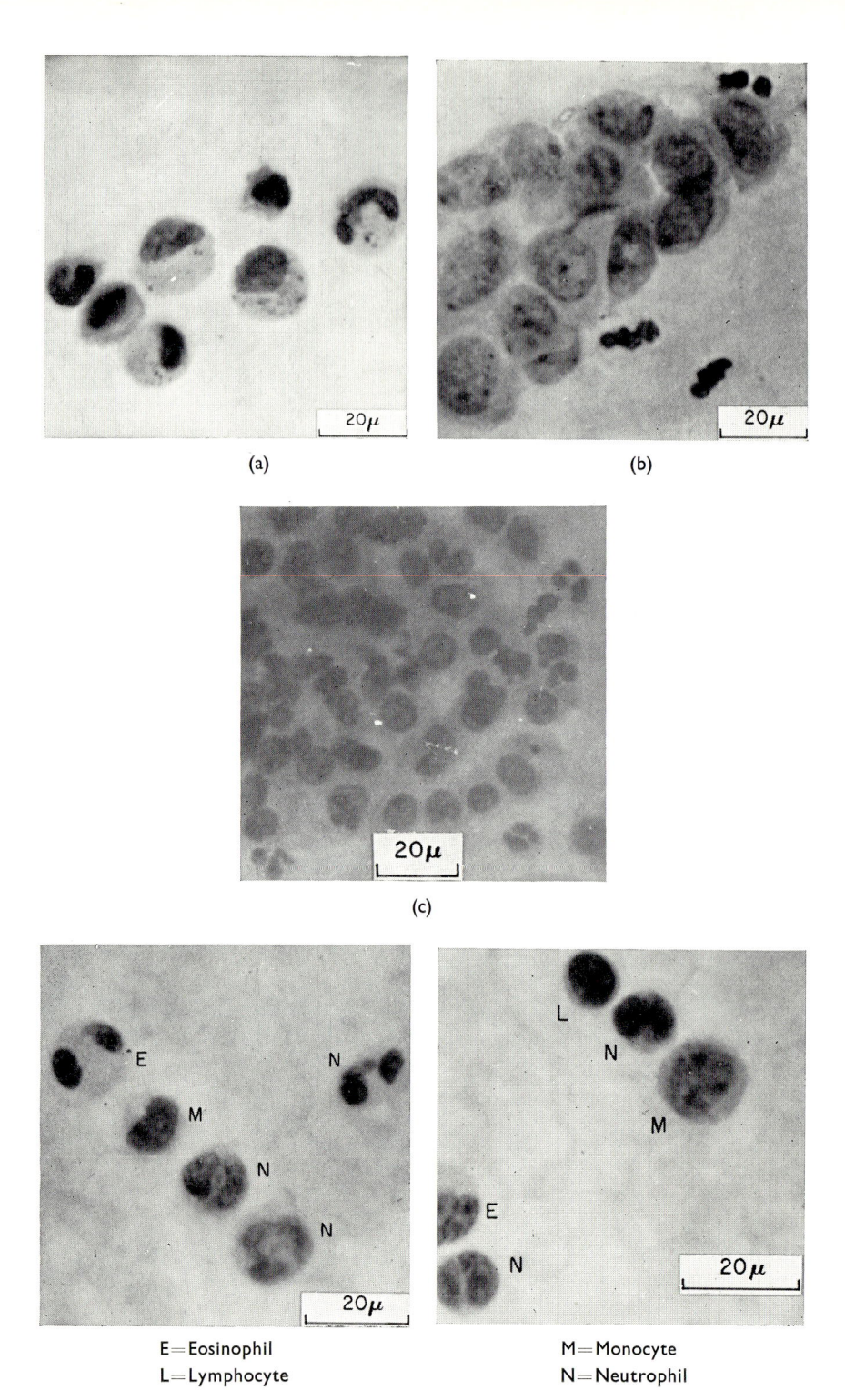

(b)

(c)

E = Eosinophil
L = Lymphocyte

M = Monocyte
N = Neutrophil

Fig. 31. (a), (b) and (c) show histiocytes.

Plasma Cells

These cells (see Fig. 32b), which are of fairly constant morphology, approximate in size to the neutrophil (12 to 14 μ). The eccentrically placed nucleus shows a chromatin pattern described as "cartwheel". Small chromatin masses are arranged in a regular radial fashion about the centre, giving the nucleus the appearance of a spoked wheel. The cytoplasm is homogenous and stains a translucent pink with the Papanicolaou stain: usually the limiting membrane of the cell is indistinct.

Plasma cells are indicative of a chronic inflammatory process.

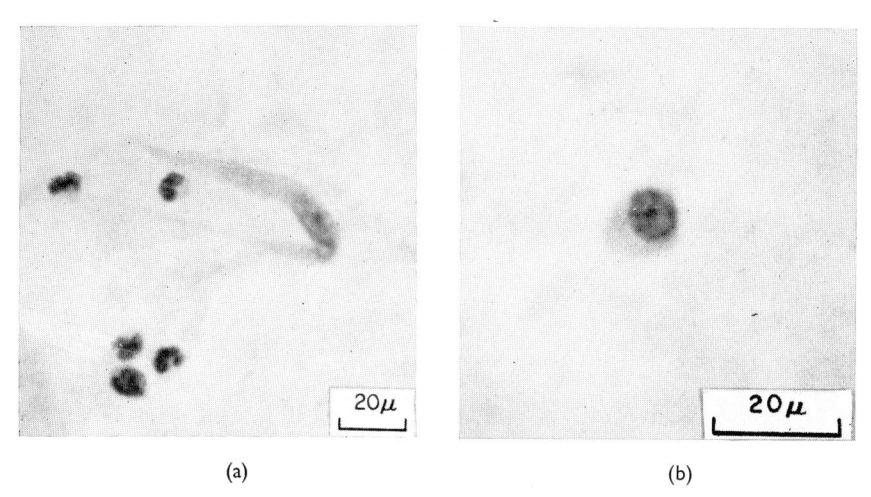

(a) (b)

FIG. 32. (a), fibroblast; (b) plasma cell.

Fibroblasts

These cells (see Fig. 32a) can exhibit considerable pleomorphism, but are often seen as long, thin cells, with pale-staining, oval nuclei possessing a finely granular chromatin pattern. They are sometimes seen in smears taken from a healing surface, and, if present in conjunction with squamous cells showing evidence of inflammatory change, circumstances may combine to bring the identity of the cells into question. In this context, the cell is long (up to 80 μ) and thin (6 to 8 μ across), with an open nucleus drawn into a sausage shape: the cytoplasm stains a pale pink.

These criteria distinguish it from the malignant "fibre cell" that may occur in smears from a squamous carcinoma.

Lymphocytes

Two types of lymphocyte may be seen (Fig. 31). The larger (about 12 μ diameter) is regarded as a juvenile form of the more frequently seen smaller cell, which is about 8 μ in diameter. The nucleus of the large lymphocyte is not completely regular in shape or texture. It is approximately round, although at times it may resemble a histiocyte in form, including the presence of an indentation. The lymphocyte nucleus stains more intensely than that of the histiocyte, so that the irregularities in the chromatin are not thrown into such sharp relief as in the histiocyte. With Papanicolaou's stain, the cytoplasm is not clearly seen; it usually appears as an indefinite, unstained or very pale pink haze that does not completely surround the nucleus.

The small lymphocyte stains much more intensely and is more regular in shape than the large lymphocyte. The nucleus is roughly round with a flattened portion on its circumference, about 5 μ across. The chromatin pattern is irregular and several irregular light-staining areas are seen. As with the large lymphocyte the cell border is indistinct, but the cytoplasm is more easily seen. It appears as a pink tag contiguous with the flattened part of the nuclear membrane, but not completely surrounding the nucleus.

These cells are associated with chronic inflammatory processes. A decisive attempt to establish personal criteria for their identification must be made, otherwise difficulty may be encountered in differentiating them from endometrial cells in gynaecological smears and the malignant "oat cells" in sputa.

Neutrophils

The ubiquitous nature of these cells (see Fig. 31) in body tissues ensures that almost every smear contains at least one or two examples. A normal, mature neutrophil is about 12 μ in diameter; its nucleus comprises a series of lobes—usually three or four—linked by threads of chromatin. Several lighter-staining areas are present in each lobe, giving the chromatin a "blotched" appearance. The cytoplasm, which has a well defined border, does not stain intensely with Papanicolaou's stain, and is either a clear blue - green or translucent pink colour.

Many neutrophils are present in smears prepared from acute inflammatory lesions and characteristic changes in morphology may be seen. Some of the cells may contain many bacteria within the cytoplasm; the cytoplasm of others may show vacuolation. Evidence of cell death may be seen from the dissolution of the nucleus. The constituent lobes separate, lose their irregular texture and assume a spherical shape before final disintegration.

Eosinophils

Identification of these cells (see Fig. 31) is readily achieved by virtue of characteristic morphology. The cell is relatively constant in size (about 14 μ in diameter) and contains a bilobed nucleus, the two lobes being in close apposition. Each lobe contains numerous irregular dark-staining masses of chromatin. The cytoplasm stains an intense pink colour and this can, at times, show a characteristic coarse granulation.

The presence of large numbers of these cells is a significant finding.

Erythrocytes

Mature erythrocytes are anucleate and appear in the smear as regular, round cells about 7 μ in diameter, unless distortion has occurred during preparation. The central concavities are not usually seen if stains other than Romanowsky are used, except when haemoglobinization is markedly impaired. Haemoglobin present in the cell confers on it an acid-staining characteristic, e.g., red with Papanicolaou. If haemoglobin is lost, as sometimes happens during fixation, the envelope takes up the basic stain (green with Papanicolaou). Similar dual staining occurs with the Gram sequence. After alcohol - ether fixation, cells containing haemoglobin stain lilac, whereas the envelope alone stains either pale pink or not at all.

CHAPTER IV

Gynaecological Cytology

The vagina is lined by a stratified, keratinizing squamous epithelium, which extends over the ectocervix. At the external os, the epithelium changes to a simple columnar type; the point where this transition occurs is called the squamo-columnar junction. This transition boundary, however, does not form a precise circle about the os, since, in the parous cervix especially, tongues of either type of epithelium may extend into the other's territory. The junction is important, because in many cases the pre-invasive, malignant epithelial dysplastic changes (*in situ* carcinoma) are seen to originate here. Invaginations of the columnar epithelium lining the endocervix produce a series of simple tubular glands. The tall, secretory cells are interspersed with ciliate cells. At the internal os, the character of the epithelium again changes, although the cells forming the uterine epithelium, called endometrium, are still columnar. These secretory cells which form into simple tubular glands, are shorter than the endocervical cells and also have ciliate cells scattered amongst them. Connective tissue (mesoderm), called stroma, separates the glands from one another and consists of irregularly shaped cells. Cytoplasmic processes from these stromal cells unite under the epithelial cells to form a basement membrane.

Collection of Smears

The following information is included to enable the technician to appreciate the reason for the presence of extraneous cells in smears and is not given as an aid to collection. A collector operating without gynaecological training may miss a clinically diagnostic finding. Apparatus designed for self-collection, e.g., the Davis cytopipette (see Fig. 33), is not generally supported; often the resulting smears are of poor quality, rendering the report valueless. An exception occurs when the gynaecologist requests serial daily smears: these patients can then be carefully instructed in the method of collection.

Material for smears is usually obtained by a scraping or brushing motion with a suitable instrument. This procedure also dislodges cells that would not normally have been shed. Two commonly used sites for collection are: the posterior fornix of the vagina; the external os of the cervix; and, occasionally, the endometrial surface of the uterus is used (see Fig. 34).

GYNAECOLOGICAL CYTOLOGY

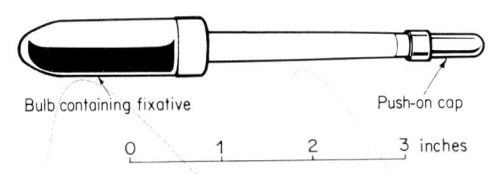

FIG. 33. Pipette for "self collection" of vaginal secretion (Davis, 1962). The fixative is injected into the vagina and then recovered in the pipette.

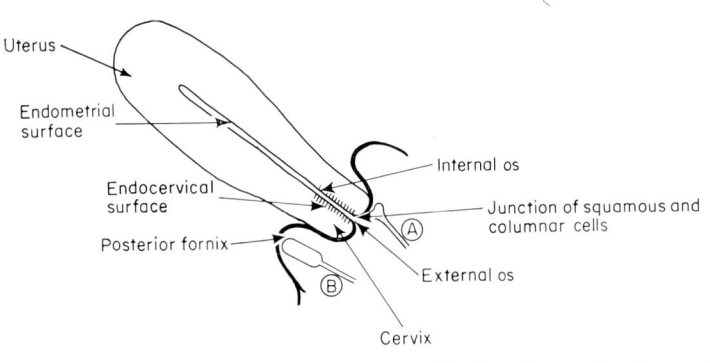

FIG. 34. Method of collecting vaginal smears with Ayre's spatula: A, cervical smear; B, vaginal-pool smear.

Vagina

Specimens collected from the posterior fornix will contain cells shed from all the epithelial surfaces (uterus, cervix and vagina) provided that the cervical canal is not occluded. The secretion is either aspirated with a special glass pipette with a stout rubber bulb or, more commonly, scooped out with the blunt end of an Ayre spatula (see Fig. 35). This instrument,

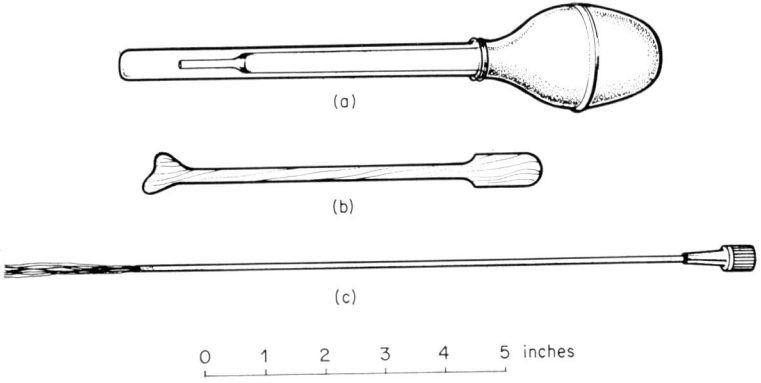

FIG. 35. Apparatus used in the collection of gynaecological specimens: (a), vaginal pipette; (b), Ayre's spatula; (c), Adam's endometrial brush.

available in wood or plastic, is cheap and disposable. Transfer of material to the slide is more difficult with the plastic type.

Cervix

The specially shaped end of the spatula is rotated round the cervix at the external os. Cells present in these smears, many of which are not exfoliated but have been scraped free, will be from the endocervix, the cervical plug and the squamo-columnar junction of the cervix. The cervix of a parous woman, so abraded, may yield blood-stained smears because the cervical epithelium is often friable as a consequence of chronic inflammation: sometimes fragments of the columnar endocervical epithelium (papillary processes) are seen in these smears.

Endometrium

Difficulties are encountered when attempting to collect cells from the endometrium directly. The technique involves passing the instrument through the cervical canal before exposing and rotating the brush (see Fig. 35). The process is not entirely without risk to the patient. During the passage of the cannula, cervical mucus containing embedded cells may be carried back, making an uncontaminated collection from the endometrial surface difficult.

Each type of smear will yield a different picture. A routine suggested is to collect separate smears from the vaginal pool and the cervix, although many cytologists request only one smear (usually from the cervix).

Macroscopic Features of the Smears

Useful information may be gathered from macroscopic examination of the smears.

Monilia Infections

The unstained smear has a dense white appearance. It is characterized by its irregular flaky nature. The flakes are plaques of squamous cells that become detached from the epithelial surface due to penetration by the hyphae that ramify beneath the superficial cells. This characteristic is completely masked if a Trichomonas infection is also present.

Trichomonas Infection

The unstained smear also has a dense white appearance. Its texture, however, is smooth and closely resembles the "rouleaux effect" seen in whole-blood films when high concentrations of globulins are present. The Papanicolaou-stained smear is characterized by the large amount of nuclear protein that has stained with the haematoxylin. This blue overshadows the largely cyanophilic elements stained by the Trichrome, giving the film a blue - green colour.

Pregnancy

After the first month and until immediately before the onset of labour the normal pregnancy smear usually presents a characteristic appearance. The unstained film has a white granular texture that is finer and more regular than in Monilia infections. This is due to aggregates of squamous cells of about equal size, which frequently gives the stained film a green colour.

Blood-stained Smears

When blood can be observed macroscopically, subsequent microscopy may indicate the cause. Possible causes include:

(a) *Cervicitis*

Smear shows numerous columnar cells including papillary processes.

(b) *Menstruation*

Smear shows numerous endometrial cells, some in characteristic "whorls"—confirm the date of last menstrual period.

(c) *Senile vaginitis*

Smear shows evidence of hypo-oestrinism.

(d) *Polyps*

Smear may show evidence of atypical cells.

(e) *Carcinoma*

Smear shows characteristic suspicious atypical cells.

Gynaecological cytology, however, may offer no conclusive findings in many of the disorders associated with abnormal uterine bleeding.

Microscopical Examination of the Normal Smear

Vaginal Pool

A striking feature of stained smears made from the normal vaginal epithelium is the crisp appearance of the cells (see Fig. 36). There is no granularity of the background mucus that is commonly present in smears prepared from infected sites. The scanty mucus that is present is seen as transparent, pale-blue, homogenous streaks that do not obscure the cells. These are, for the most part, superficial squamous epithelial cells, which have almost entirely dissociated one from another during the preparation of

PRACTICAL CYTOLOGY

he smear. The only other cell to be seen in the normal smear is the neutrophil. Small numbers of these are regularly interspersed between the squamous cells. The epithelial cells are regular in shape and size, although the staining characteristic of the cytoplasm may be either eosinophilic

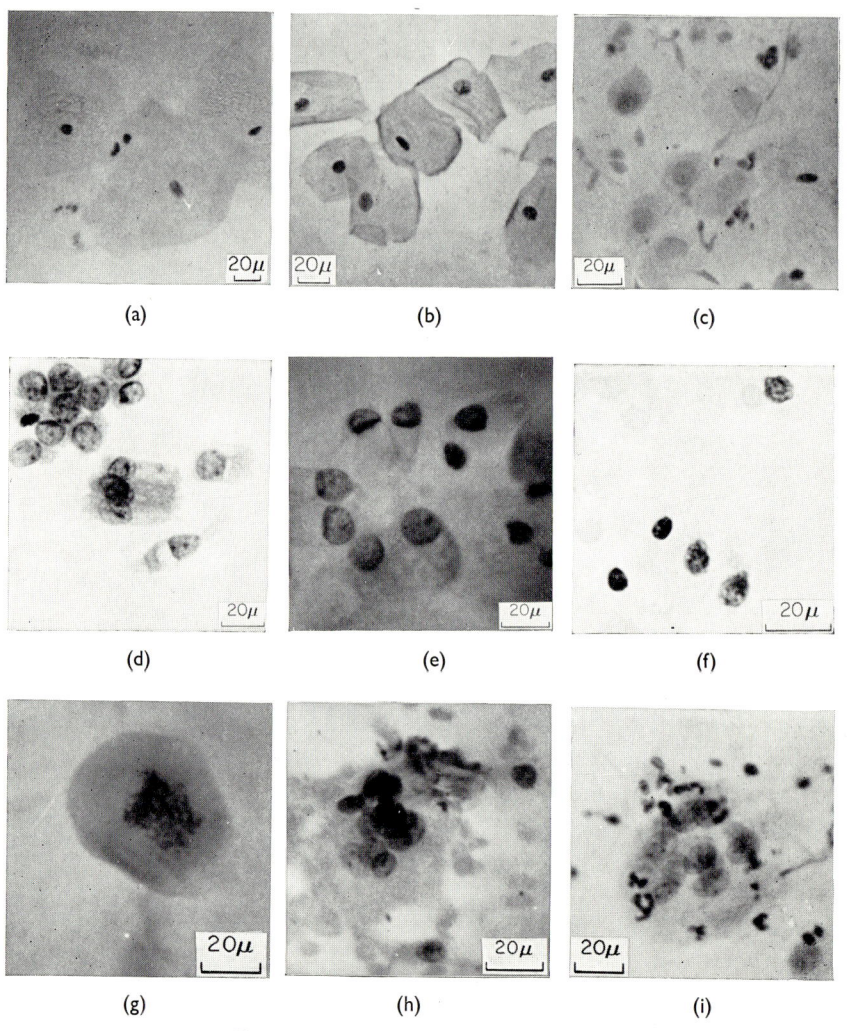

Fig. 36. (a), normal superficial squamous cells; (b), squamous cells showing progesterone effect; (c), parabasal cells from atrophic epithelium; (d), normal ciliate columnar cells; (e), normal secretory columnar cells [(d) and (e) are endocervical cells]; (f), endometrial cell; (g), (h) and (i), atypical cells seen after radiotherapy.

(staining red) or cyanophilic (staining green). All have sharply defined margins and are characteristically transparent. The ratio of the two types provides an indication of hormone activity (see p. 76). The nuclei of these squamous cells are spherical, and appear, therefore, as regular circles that vary in size from about 14 μ to 6μ (red cell diameter about 7 μ). The limiting membrane of the nucleus is contiguous with its adjacent cytoplasm. The larger nuclei have a chromatin pattern that is finely reticular and of regular density, except for the perimeter of the circle, which appears as a line of regular thickness. A small ovoid body, a sex chromosome, about 1·5 μ long may be present and appears as a densely staining mass of chromatin attached to the inner wall of the nuclear membrane (see Chromosomal Sex, p. 131).

As these cells mature, the nuclear chromatin condenses with a consequent progressive decrease in size and increase in staining intensity until at its smallest size (about 4 μ diameter) the cell is exfoliated.

Occasionally, otherwise normal smears may show a few scattered lactobacilli, which stain with the haematoxylin.

Cervical Smear

Smears from the healthy cervix are much the same as from the vaginal pool. If material from the mucus plug in the os is carried on the spatula, it will appear on the slide as threads of mucus with numerous neutrophils adherent to it. A difference in the ratio esosinophilic- to cyanophilic-staining squamous cells may exist between the smears from the cervix and vaginal pool.

No blood is seen in either of these smears from the non-menstruating, healthy woman, provided there is no trauma at collection.

The proportion of "normal" smears seen in routine gynaecological cytology is very small, possibly less than 1%: more frequently seen are smears showing evidence of an inflammatory process, usually originating in the cervix (cervicitis) (see p. 78).

Gynaecological Smears

Effect of Oestrogens on the Vaginal Epithelium

Oestrogens exert a profound influence on the genital epithelium. Cytological evidence of this ovarian activity can be seen in vaginal smears.

In the newborn, the vaginal epithelium is thick, consisting of many layers of cells. The upper stratum consists of mature, polyhedral cells with pyknotic nuclei, a condition brought about by the presence of residual maternal oestrogens. During the next few weeks these cells are shed, leaving an epithelium only a few cells thick. Vaginal smears taken during this phase, which persists until the approach of puberty, contain squamous cells of parabasal type. These cells have a cyanophilic cytoplasm and a large, open nucleus whose thin membrane encloses a fine, regular chromatin reticulum. With the approach of puberty, the vaginal epithelium again increases in thickness and at the menarche conforms to the cyclical pattern of the sexually mature woman.

Hypo-oestrinism

When sexual function reaches the climacteric, ovulation no longer occurs: the previously synchronized production of ovarian hormones becomes deranged and finally ceases. These changes are reflected in the vaginal epithelium, which, once more, becomes only a few cells deep and is referred to as "atrophic" epithelium. Smears taken during this phase, however, do not conform to a constant pattern. Not infrequently, the squamous cells present in post-menopausal smears are the normal superficial cells of the fertile woman. An explanation for this has not yet been established. In others, the normal process of maturation of nucleus and cytoplasm is disorientated. Stratification of the squamous cells no longer occurs and parabasal cells form the outermost layer of epithelium. This surface is easily disrupted so that smears contain many parabasal cells with open nuclei and translucent cytoplasm, which may be eosino- or cyanophilic. In these smears, many of the cells may be fragmented so that numerous naked nuclei are seen. The chromatin pattern of these nuclei is lost and the nucleoprotein stains as a pale lavender, usually ovoid structure.

In other smears the nuclei may show chromatin condensation. Pyknosis does not follow the normal pattern, and examples of bizarre karyorrhexis are often seen. The cytoplasm is more frequently eosinophilic and the chromatin stains blue - black. Multi-nucleate cells may also be seen.

This picture, sometimes referred to as the "atrophic" smear, is usually associated with numerous neotrophils and often blood. The administration of oestrogens causes reversion of this "atrophic" pattern to that seen during sexual maturity. A similar transition occurs during oestrogen therapy in the pre-menarche child.

Hyper-oestrinism

Although not invariably so, increases in oestrogen levels may be reflected in the vaginal smear. Such smears are characteristically "clean", i.e., free from "phage cells" and bacteria. Blood is sometimes present in variable amount, and scattered, single endometrial cells may be seen. The cytoplasm of the squamous cells is well angulated and eosinophilic, its colour ranging from pink to opaque yellow. These superficial cells, which are dissociated one from another, have small (about 4 μ diameter) round nuclei staining opaque black. Some of the nuclei may show karyorrhexis and an occasional cell may be anucleate.

"Oestrone" Grading

An arbitrary classification of smears has been practised in this laboratory as a means of indicating oestrone activity. Smears are designated one of five grades (see also Fig. 52, p. 106):

"Oestrone" Grade	Interpretation
A	Increased
B	Normal
C	Slight reduction
D	Impaired
E	Grossly impaired

The principal criterion used is the average nuclear size of the squamous cells with angularity of the cytoplasm as a secondary standard. A normal Grade B smear contains representative cells from the superficial layers only; their nuclei, vesicular and condensed, fall within the range 12 to 5 μ in diameter. Smears in which all the squamous cells possess pyknotic nuclei 5 μ in diameter or less are regarded as indicative of hyper-oestrinism and placed in category A. Grade C smears contain a proportion of intermediate cells, and Grade D smears contain predominantly intermediate and parabasal cells. Smears that contain mainly parabasal cells—normal or abnormal—are assumed to indicate hypo-oestrinism and are classed as Grade E.

Changes during the Menstrual Cycle

In the normal, fertile woman the cyclical variation in hormone levels associated with ovulation is reflected in the vaginal squamous cells. Progesterone stimulates their proliferation and maturation as far as the intermediate cell stage, and subsequent development into stratified superficial cells is directly influenced by the presence of oestrogens. By using daily differential counts, a general pattern of change can be demonstrated, but accurate counts are not easy technically to obtain. Random variation, due to methods of collecting material, preparing the smears, and counting the cells, are largely beyond control. It is not possible, therefore, to obtain much useful information about a patient by studying a single smear. Examination of a series of smears, preferably daily, during one or more complete cycles enables more definite conclusions to be drawn.

(a) *Collection of smears*

Smears for these investigations are usually prepared by scraping the wall of the upper third of the vaginal vault.

Traumatic desquamation, induced, for example, by recent intercourse, douches or inflammatory agents, will invalidate results.

(b) *Indices*

Three indices have been evolved to aid assessment of hormonal activity. These are:

1. The cornification index, sometimes called the karyopyknotic index. In the absence of an agreed standard index, the cornification index is most frequently used in this country.
2. The maturation index.
3. The eosinophilic index.

(*i*) *Cornification index (CI)*. Each squamous cell is placed in one of two classes in respect of nuclear pyknosis. The proportion of cells with pyknotic nuclei is expressed as a percentage of the total seen.

Published reports show that a considerable divergence of opinion exists concerning the definition of a squamous cell with a pyknotic nucleus. Workers returning the highest CI values are those whose criterion lies in the texture of the chromatin, taking no account of nuclear size. Such a "cornified" cell has chromatin whose open character has just disappeared, giving the nucleus a uniform opaque blue appearance. The diameter of the nucleus may be as large as 10 μ. Further condensation of the chromatin occurs until the nucleus is 4 to 5 μ in diameter and now staining opaque blue - black before the cell is exfoliated. Counts, therefore, that include nuclear size as an additional criterion will yield much lower values for the same smear.

(*ii*) *Maturation index (MI)*. Each squamous cell seen is catalogued under one of three morphological types: parabasal; intermediate; or superficial.

When a representative number of cells (see later) have been classified the result is expressed as a differential count in a standard form 0/0/0. For example, 2/38/60 indicates that the count is parabasal cells, 2%; intermediate cells, 38%; superficial cells, 60%. On the rare occasions when basal cells are seen, they are included in the parabasal cell count. A progressive increase in the percentage of superficial cells is described as a "shift to the right", and the converse as a "shift to the left".

(*iii*) *Eosinophilic index (EI)*. A differential count of the squamous cells is made by classifying them according to the staining characteristic of the cytoplasm. The number of eosinophilic cells is expressed as a percentage of the total number counted.

(iv) *Reproducibility of indices.* Many random factors affect the accuracy of these counts and can make interpretation difficult. The epithelial surface should not be traumatized during the collection of the smear and the material must be evenly spread over the slides. Many cells must be classified before the index can reflect the trend of the squamous cells. The actual number to be counted varies considerably from slide to slide, but generally a minimum of 500, taken from the total area of the smear, is required. To ensure correlation of results, the serial counts for a particular patient should preferably be carried out by the same observer.

(v) *Results.* In normal, fertile women the curve of daily plasma oestradiol levels shows a peak at ovulation with a secondary smaller peak in the mid-secretory phase (see Fig. 37). Progesterone levels are steady until the secretory phase, when the rising curve is maximal coincident with the secondary oestradiol peak. The normal pattern emerging from the differential counts is of a gradual progressive rise that is maximal at ovulation: thereafter the curve falls, returning to its base-line at menstruation, thus roughly following the oestradiol curve. A characteristic "progesterone effect" is seen in the squamous cells during the secretory phase. Curling of the cytoplasm occurs and produces a typical "wavy-edge" appearance. Some of the superficial cells have parts of the cytoplasm folded back over the cell ("envelope effect"). Intermediate cells, similarly influenced exhibit a typical boat-shaped appearance and are termed "navicular cells". The transparent cytoplasm becomes translucent and granular, and there is a progressive increase in the number of cyanophilic cells, which tend to remain clumped in groups.

The description of the "normal" vaginal smear described earlier, refers to that taken from the normal fertile women at mid-cycle, i.e., ovulation. As the cycle progresses, the changes in the squamous cells outlined above take place. After ovulation, in the absence of conception, the number of

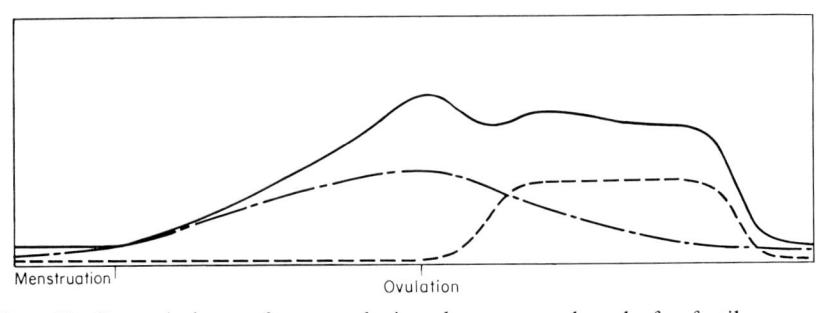

FIG. 37. General shape of curves during the menstrual cycle for fertile, normal women: ———, oestradiol; — — — —, progesterone; — · — · —, cytology indices.

neutrophils and histiocytes progressively increases until just before menstruation. At this stage smears present what has been called "a dirty appearance". Numerous neutrophils and histiocytes with fragmentary cells and many bacteria are seen.

At menstruation, groups of characteristically whorled, endometrial cells are present. Single endometrial cells are often partly obscured by the background of blood. The chromatin pattern of the endometrial cells generally shows gross irregularity in texture and staining properties; neutrophils may be seen in the cytoplasm of some of the cells.

Pregnancy

After the first month and until just before term smears taken during a normal pregnancy have the characteristic macroscopic appearance already described. When stained, the smears are uniformly blue - green in colour.

Microscopy shows that the granular appearance is due to squamous cell aggregates of regular size—this has been referred to as "plaquarding"—with only a few neutrophils. The squamous cells are intermediate "navicular cell" forms with only an occasional superficial cell present. The cyanophilia and folding are manifestations of the markedly increased progesterone levels present during pregnancy and parallel the progesterone effect seen in post-ovulatory smears. The relatively low oestrone levels prevent condensation of the nuclei, which are, therefore, open in texture with a regular nuclear membrane. During pregnancy cytoplasmic inclusions in the squamous cells may appear. These are spherical masses, one or many, that vary in colour between golden yellow and black. Smears may also show evidence of cytolysis accompanied by numerous lacto-bacilli, but with no increase of neutrophils.

Neither of these findings, however, specifically identify a "pregnancy smear".

Just before labour, the picture alters and plaquarding disappears. Many of the discrete cells are of the superficial type, although the cyanophilia usually persists.

After delivery the squamous cells show the typical picture of hypo-oestrinism, which continues during the period of lactation. The smear pattern returns to its normal cyclical nature when lactation ceases.

The "Inflammatory" Smear

Many of the smears collected from women attending gynaecological clinics and "well women" centres will fall into this category. In most instances this is due to a chronic cervicitis. This can vary from infection and slight degeneration to gross ulceration. An infected cervix yields a blood-stained smear with many neutrophils and histiocytes. Numerous types of bacteria may be present and a Gram film may show actual colonies, together with neutrophils exhibiting intra-cellular organisms. Epithelial cells show the inflammatory changes already described. Inflammatory smears are also obtained in cases of vaginitis, e.g., trichomoniasis and moniliasis.

General features

Increased numbers of intermediate and parabasal cells are present and, not infrequently, some of these cells exhibit multi-nucleation. Plaques of epithelial cells infiltrated by neutrophils are often seen. In some smears the epithelial cells may be seen to possess intracellular neutrophils. A distinction between these and neoplastic cells, which may also contain neutrophils, is that "inflammatory" cells may contain as many as twenty or more neutrophils within a single cell, whereas neoplastic cells generally contain only one or two.

Columnar cells, from the endocervical glands, may also be seen. These may be present singly or in small groups showing changes described under Atypical Epithelial Cells (p. 46), or as fragments of tissue - papillary processes. The groups are easily identified by virtue of the regular architecture of replicate cells exhibiting a typical "honey-comb" structure (see Fig. 38). On focusing at the periphery of these aggregates, a regular line of normal columnar cells is seen. Other plaques of these cells may show the metaplastic change also described under Atypical Epithelial Cells.

Moniliasis

Changes, more or less characteristic, are seen in smears taken from epithelia colonized by this pathogenic fungus. This infection is found in 2·3% of the gynaecological smears examined in this Unit: a further 2·1% of smears contain budding cells without hyphae. In the absence of a Gram-stained film, the fungus can very easily be missed.

Identification

In its vegetative form the fungus consists of long, branched filaments called hyphae (see Figs. 39 and 40), which generally stain Gram-positive. At times they may stain indifferently and occasionally are frankly Gram-negative. The growing hyphae frequently show the presence of buds, which are the spores of the fungus. These stain uniformly Gram-positive, and when large numbers are present in the smear no hyphae are seen. Spores may multiply by "budding"; a small outgrowth set at an angle to the long axis of the oval cell. This gradually increases in size before itself producing a bud.

The fungus is sometimes missed in Papanicolaou-stained smears, because of its indifferent, pale staining characteristic. Suspicion may be

80   PRACTICAL CYTOLOGY

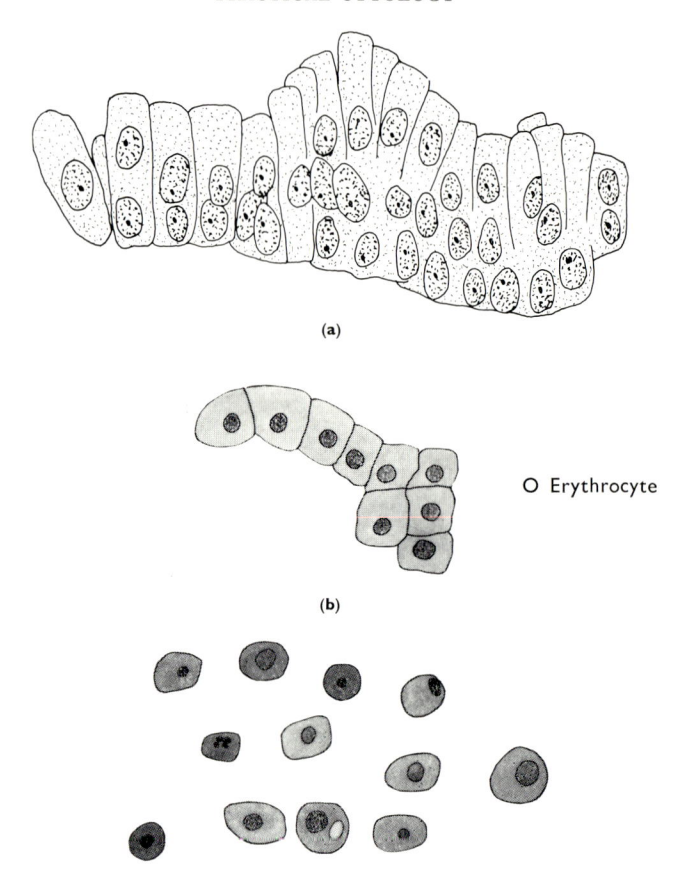

(a)

(b)   ○ Erythrocyte

Fig. 38. Columnar cells showing: (a), inflammatory change; (b) metaplasia.

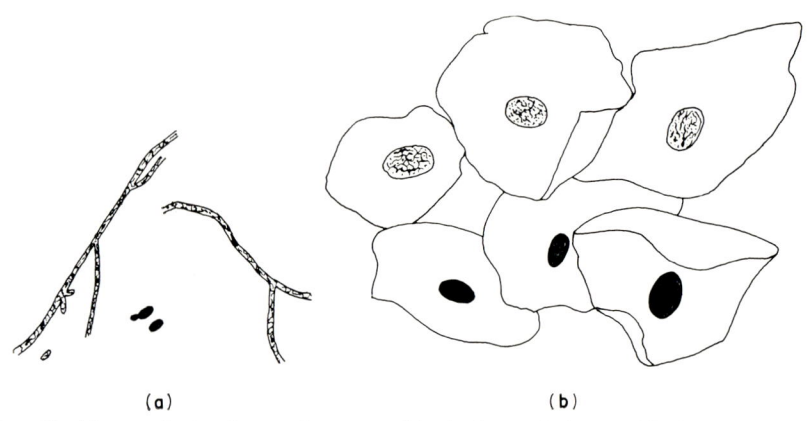

(a)   (b)

Fig. 39. (a), monila hyphae and spores; (b), dyskaryosis in moniliasis—note ovoid nuclei (see also Fig. 26e).

aroused, however, by the clumping and changes seen in the squamous epithelial cells.

The flaky macroscopic appearance has already been described. Microscopy shows the flakes to consist of plaques of superficial squamous cells, and between the plaques are dissociated superficial, intermediate and parabasal cells. The nuclei of these cells frequently exhibit a swollen, ovoid appearance, and the texture of the chromatin is regular and finely granular

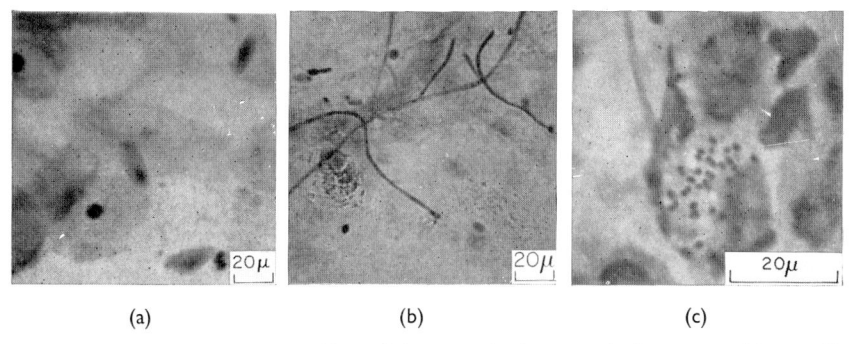

(a)   (b)   (c)

FIG. 40. Vaginal pathogens: (a), trichomonads in a typical smear; (b) monila infection; (c), gonococcal infection.

Some of the superficial cells, particularly in association with the plaques, are dyskaryotic. They are larger than normal and their nuclei proportionately large, are opaque, deep blue or black. An ill-defined perinuclear blue haze is sometimes seen in the cytoplasm, which otherwise stains an opaque yellow.

Trichomonas Vaginalis

This flagellate protozoan follows a parasitic existence in the genital tract.

Microscopy of saline preparations made from the foamy, creamy vaginal exudate in trichomoniasis may demonstrate the living organism. In these circumstances identification is relatively easy; agitation of cells in the microscope field alerts the observer to scrutinize with a higher magnification when the pear-shaped organism with its oscillating flagella is quickly seen. Direct microscopy of prostatic secretion or urine deposits obtained from the consorts of infected women is less successful, although culture techniques may succeed. Failure to recognize the dead organism in saline preparation is sometimes suggested to explain anomalies between direct examination and culture results.

In Papanicolaou-stained smears, trichomonads exhibit a wide variation in form and staining characteristics. This can confuse an inexperienced worker. Most of the criteria used for identification in the saline preparation are valueless when examining smears. Once experience is gained, however, statistics show a higher incidence of positives in Papanicolaou-stained smears than in wet preparations. Analysis of results from 20,000 routine gynaecological patients in this Unit reveals an incidence of about 8%. Further examination of results shows that in smears from female patients attending venereal-disease clinics, the incidence approaches 40%, whereas in women attending "well women" clinics the figure is about 4%.

Identification

In smears trichomonads appear as irregularly shaped cells with ill-defined borders (see Figs. 40 and 41). The flagella are only occasionally seen. Although the cells are frequently smudged, intact trichomonads are generally ovoid and vary between 7 μ and 30 μ long. A nucleus, sometimes described as almond shaped, but frequently round, 2 to 3 μ in diameter, is seen on the central axis situated toward the blunt end. The translucent cytoplasm may stain red, green or lavender with Papanicolaou. Red inclusions may be seen in the cytoplasm of specimens staining green or lavender. Identification rests on the presence of the nucleus in the cell. Usually there are numerous examples in positive smears.

Dyskaryosis of the epithelial cells is a common feature and is most noticeable in the superficial squamous cells. Numerous larger-than-normal superficial cells are frequently seen with opaque yellow or red cytoplasm and dark-blue, structureless nuclei. Similar abnormalities are seen less frequently in the intermediate and parabasal cells that are usually present in these smears. Other squamous cells may show an enlarged, open nucleus, with a well defined nuclear rim. The contained chromatin precipitates in an irregular fuzzy haze, which gives the chromatin a skeining effect.

Irregular precipitation of the nuclear chromatin of all the cells in some smears is a common observation. The general appearance of both epithelial and non-epithelial cell nuclei is of a nucleus with a clearly seen, regular nuclear membrane and a central cleared area.

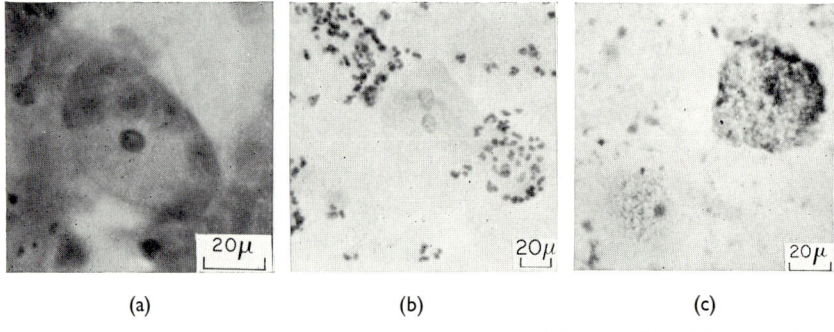

FIG. 41. Possible stages in trichomonal vaginitis: (a), juvenile trichomonads on superficial cell; (b), neutrophils adherent to superficial cells; (c), bacteria coating a superficial cell (Gram's stain).

Another common feature is the presence of partial or perinuclear haloes in the epithelial cells. The pyknotic nuclei of affected superficial cells are thus thrown into sharp relief.

Suspicion of trichomoniasis may be aroused by the following secondary features:

1. Increased numbers of neutrophils are almost always present in this infection, although occasionally the flagellate may be seen without pus. The neutrophils are spread evenly through the smear, except for an occasional tightly packed cluster. Examination with a ×40 objective shows that this is caused by neutrophils adherent to the surface of a single superficial squamous cell. Close scrutiny of such a cell, partly free from neutrophils, discloses numerous pits in its surface. Less frequently, careful examination of a film will show an occasional superficial squamous cell with juvenile trichomonads, 8–10, colonizing the surface. Close scrutiny of the exposed part of the surface of such a cell will also show numerous pits.
2. A pleomorphic Gram-variable coccobacillus is always present in association with trichomoniasis. As the density of this organism in the smear increases, the number of trichomonads seen decreases. When the organism is in abundance, no trichomonads are seen. The organisms are spread evenly through the film, except for an occasional tight cluster. Scrutiny with the ×40 objective shows that this is due to a superficial squamous cell, that is heavily coated with the organism.
3. The leptothrix associated with trichomoniasis (Koss and Durfee, 1961) has been much less frequently identified in this Unit (about 0·1%) than the organism described above.

Suspicion of trichomoniasis arising from these secondary features may or may not be confirmed by the finding of trichomonads in the smear.

The "Malignant Smear"

In gynaecological smears, the morphology that malignant epithelial cells exhibit ranges so widely that a confident assertion of cell types likely to be encountered in a positive smear is not possible. While remembering the hazards of generalization, the most outstanding atypical features are anisonucleosis and nuclear anisochromia, which are sufficiently striking to alert the experienced observer scanning at ×100 magnification. Examination at ×400 magnification reveals more detail concerning the variation in chromatin texture of the nuclei of affected cells and also the deviation from the normal nucleus/cytoplasm ratio becomes more apparent.

Some caution should be exercised in deciding the type of epithelial cell involved when malignancy is established on purely cytological grounds.

Not infrequently, the malignant cells arising from a squamous lesion exhibit characteristics more in keeping with aberrant columnar cells, and, conversely (but less commonly perhaps), the cells from an adenocarcinoma masquerade as atypical squamous cells (see Fig. 42). Generally speaking, however, smears from pre-invasive and early invasive squamous lesions contain numerous examples of markedly atypical cells that are manifestly squamoid in type. In other smears, especially from adenocarcinomata, the typing of the cells may be more difficult, although the atypical cells can usually be related to one another through secondary cytoplasmic characteristics, despite gross dissimilarity of their nuclei. When the atypical cells possess minimal cytoplasm, differentiation of their nuclei from non-malignant "naked nuclei" must be based on chromatin texture.

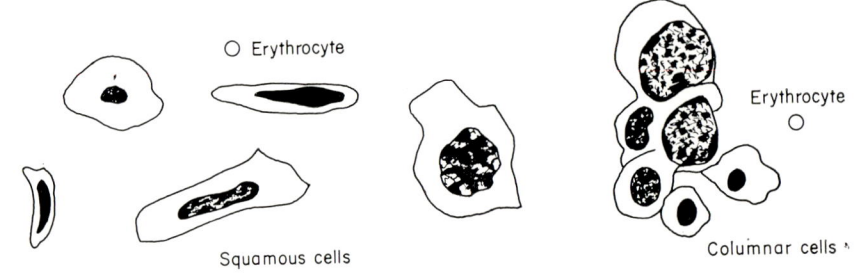

FIG. 42. Illustration of the range of anisonucleosis and nuclear anisochromia likely to be encountered in "malignant smears".

Too close an application of the criteria for malignancy (see p. 47) may result in indecision, but it should be remembered that the Papanicolaou slide grading is not a rigid classification of cells. It is a personal assessment based on experience, for which there is, as yet, no substitute.

Radiation Changes

X-rays and other high-energy emissions damage living cells. Neoplastic epithelial cells are generally more sensitive to irradiation than are their normal counterparts. As a consequence of irradiation, many cells become non-viable and the morphology of the subsequent progeny of surviving atypical cells is unpredictable. Many of the new cell-lines quickly disappear; others, which may persist in irradiated tissue for years, exhibit bizarre morphology.

(a) *General features of the smear*

Post-irradiation smears are usually heavily blood-stained, with numerous neutrophils and histiocytes present. Changes occur in both nucleus and cytoplasm of epithelial cells: many exhibit marked increase in size. Nuclei may show abnormal chromatin condensation, *viz.*, pyknosis, karyorrhexis and karyolysis: others appear crenated. The cytoplasm is non-angulated, usually eosinophilic and vacuolated. These may be large, single vacuoles, situated at the periphery or a series of small irregular fenestrations. The number of bizarre cells in the smear drops with the passage of time, but one type that often persists is a giant multi-nucleate cell, in which nuclear replication without division of the cytoplasm occurs (see Fig. 36). Most of the epithelial cells seen, however, are parabasal cells and a general impression of most smears is of a blood-stained "oestrone deficient" film.

Attempts have been made to use cytological methods as a means of predicting the susceptibility of malignant cells to radiotherapy—(the sensitivity-response (SR) test) and assessing the effectiveness of such treatment (the radiation-response (RR) test).

(b) *Sensitivity-response test*

In this test (Graham and Graham, 1953), the parabasal cells, in a smear collected before treatment, are examined, and the proportion of cyanophilic cells with vacuolated cytoplasm is expressed as a percentage of the total. Patients whose smears yield 10% or more SR cells are said to benefit from radiotherapy.

(c) *Radiation-response test*

In this test (Graham and Graham, 1955), daily smears are taken, commencing immediately after irradiation. The radiation response is assessed by estimating the percentage of benign cells showing radiation changes. At the peak, counts of 75% or more suggest a good response; less than 60% is adjudged a poor response.

CHAPTER V

Sputa

Interest in the increased incidence of carcinoma of lung has brought renewed attention to the use of cytology as an ancillary aid to early diagnosis. It has been suggested (Long, 1963) that individuals at risk, heavy smokers for example, should submit sputum specimens at regular intervals for examination.

Collection of Material

Sputum collected without safeguard may contain material (cells, mucus and foreign matter) from any of the epithelial surfaces of the respiratory tract, including those lining the buccal cavity. Some precaution can be observed in order to reduce this contamination. Sputum produced by a fasting patient following a deep cough after first rinsing the mouth with water produces a satisfactory specimen. Various irritant aerosols have been designed to induce a violent cough and so dislodge deep-seated mucus.

It is usual to receive a series of three of these early-morning specimens. This is a minimal requirement. Even if the first specimen appears to yield adequate evidence of malignancy, the rest of the series should be examined in order to eliminate any possibility of error, either in diagnosis or processing.

Several workers have suggested that it is desirable to pass sputum directly into fixative. Alcohol, ethanol or isopropanol, in concentration ranging from 50% to 70% in water is usually recommended: at these concentrations mucoprotein is not coagulated. For most of these methods, the fixation time is given as 30 min. In this Unit, specimens are collected in dry, sterile polythene containers. Sputum has been found to keep for quite long periods, up to 12 h at room temperature and for several days at 4°C, without impairing the quality of subsequent smears. In warm weather, specimens begin to deteriorate after 4 or 5 h at room temperature. The first cytological evidence of this is autolysis of the neutrophils, followed by similar changes in the histiocytes; both the normal and abnormal epithelial cells appear unchanged. It is quite acceptable, therefore, to receive sputa by post.

During the clinical investigation of chest lesions, a catheter may be passed into a bronchus. This allows the collection of cells from a particular area of lung without contamination from other respiratory surfaces. Cells are obtained by passing small volumes of isotonic saline through the catheter by a syringe. The saline suspension of cells is then recovered and sent to the laboratory for processing.

Preparation of Smears

Sputa can be processed in two principal ways: (*a*) by preparing smears directly; or (*b*) by embedding the specimen in wax and sectioning. Either method can be applied with or without a preliminary concentration technique. Concentration procedures are said to yield more accurate results more quickly than smears prepared directly. In deciding which method of preparation to adopt, it should be remembered that concentration involves the risk of distortion or disintegration of cells, particularly of neoplastic cells, and also that supplementary information concerning proportionate numbers of the types of cells present is destroyed. The histological technique is dealt with in the section on Ancillary Methods (see Rapid Wax Embedding, p. 116).

A "direct smear" technique is used in this laboratory, careful attention being given to the selection of material for preparing the smear. The unfixed specimen is transferred to a disposable Petri dish. Two hardwood applicators are taken and one end of each is frayed. This can be carried out by crushing one end between scissors or simply snapping off the last half-centimetre to leave a ragged end. The applicators are used to spread the sputum over the total area of the plate. The specimen is then scrutinized; a magnifying glass ($\times 10$) or a plate microscope is a considerable help at this stage.

Points to look for are: white flakes or thin white threads; tiny black specks; and flecks of blood.

White flakes and threads are usually indicative of either normal or abnormal epithelial cells.

Black specks are due to the presence of carbon-containing histiocytes—a pointer to a satisfactory specimen.

Blood flecks are usually present in association with atypical (not necessarily malignant) cells.

If the specimen is shown to consist of only a thin clear fluid, it may be discarded as unsatisfactory. It is extremely important, however, to make quite sure that no thin white threads or tiny black specks have been overlooked, since such specimens may contain malignant cells.

When flecks of blood are present they should be taken for examination.

Sputa that are purulent or heavily blood-stained should be examined for the black specks, and these parts should be selected for microscopy.

Parts of the specimen exhibiting these characteristics can be removed quite easily with the frayed ends of the applicators. Smears are made from the whole of the selected fragments and fixed in alcohol - ether solution before drying occurs. It is helpful to prepare at least one other smear and to fix in methanol. This smear is stained with a Romanowsky sequence.

The following concentration techniques illustrate the general pattern of most of the published methods. It should be stated that all claim a higher proportion of confirmed positive smears than the direct-smear technique. This may well be the case for an experienced cytologist, but a beginner is well advised, initially, to examine a direct smear preparation in association with smears prepared by concentration techniques.

Saccomano and Saunders (1963)

The patient is instructed to wash out the mouth with water. A "deep-cough" specimen is then collected into a container half-filled with fixative. This comprises 50% alcohol with 2% polyethylene glycol in water, and is said to prevent shrinkage of the cells. The specimen is mixed by shaking and can be left for several days before examination.

To prepare the smears, the specimen is homogenized in a blender for 30 sec and then centrifuged at 1,500 rev/min for 5 min. Smears from the deposit are made and fixed immediately. It is claimed that this treatment will not break up plaques of cells.

Abramson, Dzenis and Hicks (1964)

A deep-cough specimen is collected in 70% alcohol and allowed to stand for 30 min at room temperature. An equal volume of alcohol - xylene solution (50 - 50) is now added and the mixture left for 5 min. Five millilitres of the solution are then centrifuged at 2,500 rev/min for 45 min, and smears are prepared from the deposit.

The paper details the preparation of paraffin-wax centrifuge tubes and the processing of the supernatant mucus.

Cells seen in Sputa

Sputum may contain epithelial cells from all the surfaces extending from the mouth to the lung alveoli (see Fig. 43). A properly collected specimen eliminates the preponderance of squamous cells and flora from the buccal cavity which, otherwise, can make examination difficult.

Stratified squamous cells line the buccal cavity, pharynx and part of the larynx. Pseudo-stratified columnar cells, secretory and ciliate, line the larynx, trachea, bronchi and broncioles, and simple squamous epithelium lines the alveolar spaces.

The squamous cells from the buccal mucosa seen in sputum are typical keratinizing, superficial cells with an occasional intermediate cell. Columnar

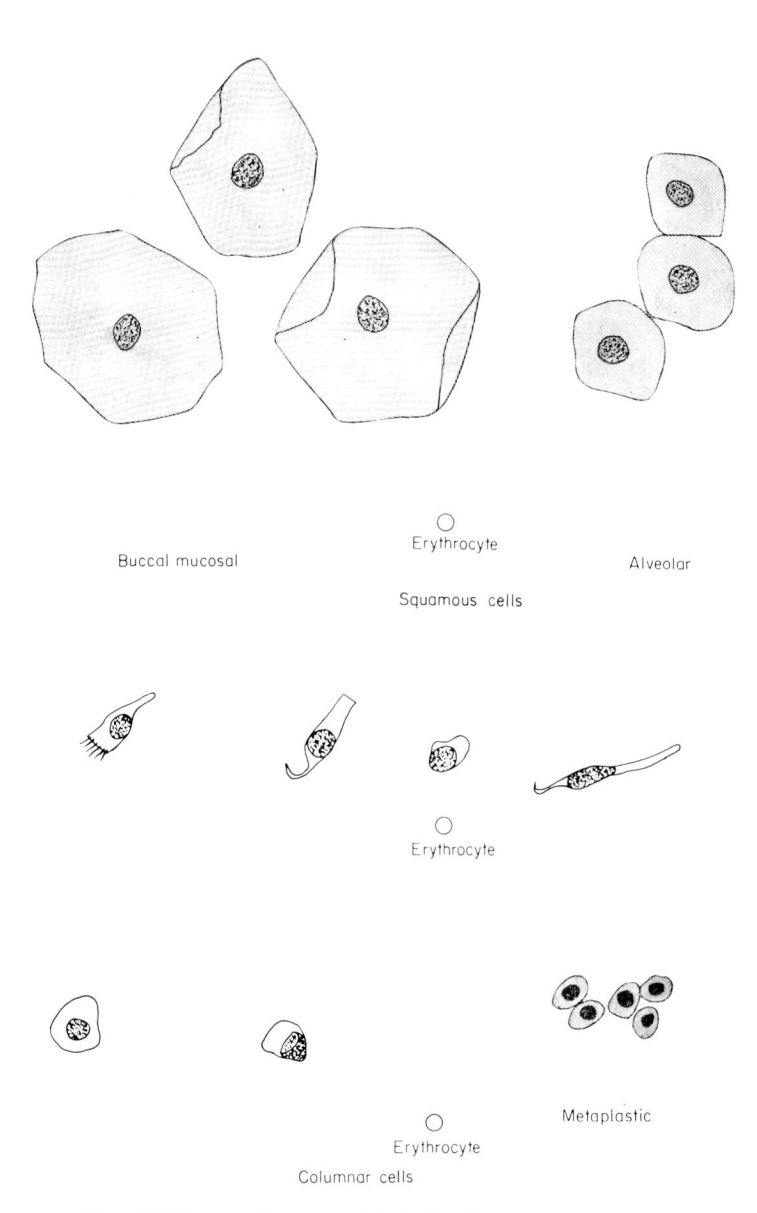

Fig. 43. Non-malignant epithelial cells seen in sputum.

cells, usually ciliate, should be seen in a properly collected specimen, although not in large numbers. In normal sputum the alveolar cells are not present. The mucus present in the respiratory passages normally contains small numbers of histiocytes, often with carbon inclusions and a few neutrophils.

A smear from a normal sputum will, therefore, contain a few squamous cells, scanty columnar (usually ciliate) cells, with small numbers of histiocytes and an occasional neutrophil. All the cells are well preserved and stain characteristically. The mucus in the smear appears as transparent pale blue streaks; the few organisms present are usually cocci.

Atypical Cells

Associated with the increased number of neutrophils and histiocytes present in respiratory-tract infections are changes in the morphology and numbers of the columnar cells. Often fragments of columnar epithelial tissue are present with varying numbers of free cells, usually ciliate rather than secretory. Sometimes these cells show changes resulting from inspissation of mucus in the bronchioles. Although the general morphology of the cells remains unchanged the normal staining characteristics are lost, the various cell structures taking up basic stains (green with Papanicolaou) with varying intensity. The size of these cells is almost always larger than normal. Metaplastic changes can be seen in some of the columnar cells and follow a pattern similar to that observed in gynaecological smears. The metaplastic cells seen in sputum, however, are generally slightly smaller and stain more intensely than those from the cervix. Their replicate character and presence in plaques assists in identification. Pyknosis of the centrally placed nucleus produces a characteristic, dark-blue, staining reaction, whereas the dense cytoplasm often stains a deep red. The limiting membrane of both nucleus and cytoplasm is thus sharply defined. If there is doubt concerning the non-malignant nature of these cells, the presence of adjacent cells with cilia serves to characterise the group as benign.

Sputa may contain other atypical features, such as epithelial pearls or squamous cells with intracellular neutrophils. These would, of course, merit review by a cytologist, but are not generally indicative of neoplasia.

Often the background of mucus in these smears is a translucent blue, and the presence of embedded bacteria gives a granular appearance. Inspissated mucus may produce the characteristic Curschmann spirals. In certain diseases, sputum may also contain quite large numbers of lymphocytes, and these may possibly be confused with cells of the "oat cell" type of carcinoma.

The presence of eosinophils, red cells and elastic fibres are significant abnormal findings; their presence should be reported and confirmatory opinion obtained.

Malignant Cells

Malignant cells seen in sputum may arise from differentiated squamous, adeno-, and "oat cell" carcinomata, which produce characteristic atypical cells, or from undifferentiated and metastatic lesions (see Figs. 44 and 45).

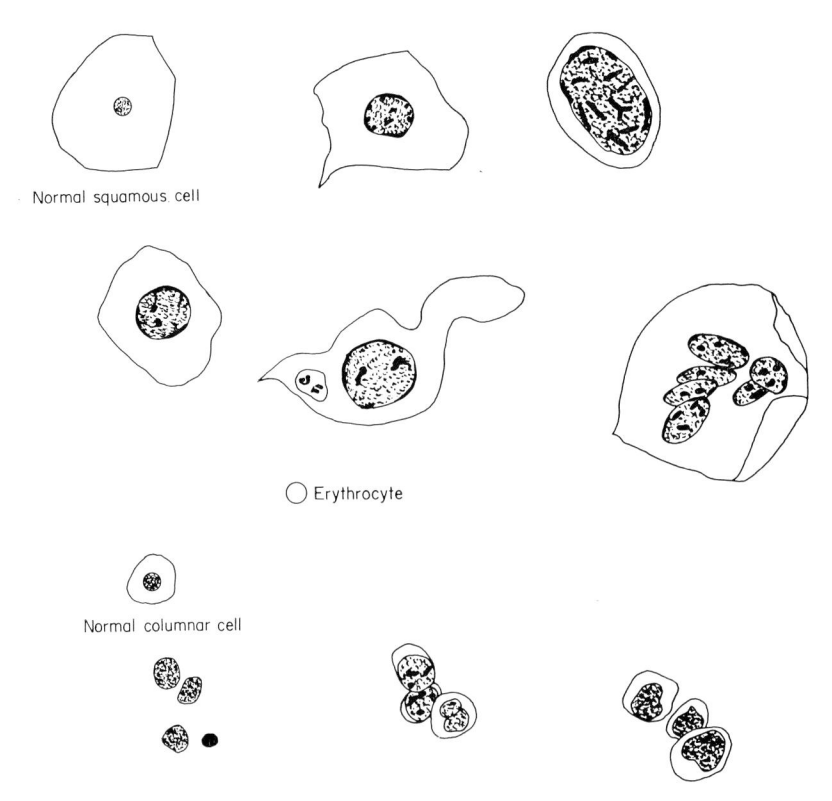

Fig. 44. Malignant cells seen in sputa.

Fig. 45. Cell group, showing "moulding".

(a) *Squamous carcinoma*

During scanning, strikingly abnormal forms of squamous cells may be seen, some of which are much larger than the normal cells present; these cells can usually be seen to be related to groups of squamous cells, which themselves show irregular morphology. The nuclei of the cells also show irregular shapes. There is uneven condensation of the nuclear chromatin, which produces a coarse, granular pattern, and the staining characteristic is almost black. The cytoplasm of single cells in the group may show a noticeable yellow colour. Multi-nucleation may be seen with a wide difference of character between the nuclei.

(b) *Adenocarcinoma*

Groups of cells are seen that show marked anisonucleosis, including a few with very large nuclei. The nuclei are usually eccentrically placed and show a coarse, open pattern, owing to irregular chromatin condensation. Nucleoli may be present staining a purple - red. The cytoplasm varies from cell to cell and may be "foamy". This latter characteristic sometimes leads to confusion with histiocytes, but this difficulty can often be resolved. Histiocytes will show a replication of the atypical characteristic, and by focusing through the cell the indentation (hof) in the nucleus can frequently be seen. Examples of the cell will also be seen showing the typical carbon inclusions in the cytoplasm. These malignant cells can easily be distinguished from normal bronchial epithelial cells, which show a constancy in morphology and staining, a fine chromatin pattern and ciliated forms that are usually part of groups exhibiting a regular architecture.

(c) *Oat-cell carcinoma*

The smear contains strings of tightly packed cells whose nuclei show "moulding", i.e., the cells in apposition are so closely impressed as to cause their nuclei to shape themselves against one another (see Fig. 45). There is only a minimum of cytoplasm, and some cells appear to possess none at all. Individual cells in the streak show nuclei that range from a dense black to a vesicular, recticular chromatin pattern. The cells resemble lymphocytes in size, but can be distinguished by their presence in streaks, the anisonucleosis and aniscochromia together with their characteristic moulding.

CHAPTER VI

Serous and Other Fluids

The walls of the pleural, peritoneal and pericardial cavities are lined with a simple squamous epithelium composed of mesothelial cells (see Epithelial Tissues, p. 34). Small amounts of serous fluid are present in these spaces; the function of the fluid is to lubricate the surfaces of contained organs. Aspiration of this fluid, which is carried out with a wide-bore needle, will, in health, yield only tiny volumes. Certain diseases, both malignant and benign, may cause abnormal accumulations of fluid to occur in the peritoneal cavity, when the condition is termed ascites, and in the pleural space, when it is called a pleural effusion. The mechanism causing the accumulation is obscure, and in those instances due to neoplastic disease, it does not necessarily follow that malignant cells are colonising the fluid in the cavity.

Preliminary Examination

The macroscopic features of the specimen should be recorded, including the volume of fluid. The total volume of aspirate should be submitted for analysis, so that all the cells are available for examination. Occasionally only a fraction of the supernatant is sent to the laboratory after the fluid has stood, for example, awaiting the completion of a request form. During this interval many of the cells will sediment and not be represented in the laboratory specimen.

Blood, if seen macroscopically, should be noted, since the preparation technique includes an attempt to eliminate red cells from the smears. "Rouleaux effect", the red cells packing together similar to stacks of coins, may indicate the presence of abnormal proteins or protein ratio. Agglutination of the red cells may be due to cold agglutinins, for example, in pleural fluids from patients whose effusion is due to a viral pneumonitis. Clots of blood and fibrin should also be noted and removed for separate examination (see Rapid Wax Embedding, p. 116). The presence of pus frequently confers a homogenous, milky appearance to the fluid. A flaky, white deposit suggests plaques or clones of cells other than neutrophils.

During the preparation of the smears the cell-free supernatant fluid obtained can be examined for approximate protein concentration with "Albustix" reagent strips (Ames Ltd), diluting with water if necessary.

Preparation for Cytology

The "cell counts" sometimes attempted on these fluids, are time-consuming, of doubtful accuracy and no value. Three methods commonly used for preparing smears are: simple centrifuging; centrifuge - flotation; and filtration. These are described under Ancillary Techniques—Concentration Methods (p. 121). Smears are prepared and fixed with: (*a*) alcohol - ether, followed by the Papanicolaou staining sequence; and (*b*) methanol, followed by Jenner-Geimsa and Unna–Pappenheim stains.

Cytology

Cells that may be present in these fluids can be divided into two groups; epithelial and non-epithelial. Group I comprises normal and abnormal mesothelial cells, from the cell-lining of the cavity, together with, for all practical purposes, any malignant cells that may be present. Group II contains the non-epithelial mesodermal cells, namely lymphocytes, plasma cells, histiocytes, neutrophils and erythrocytes.

Group I. Epithelial Cells

Mesothelial cells may be seen in smears obtained from fluid accumulations due both to neoplastic and inflammatory causes. Normal cells may be examined by preparing smears from the external surface of organs obtained by operation from, for example, the abdominal cavity. The size of the cell is generally about 25 μ in diameter, the nucleus measuring 14 μ across. The limiting membrane is round or roughly ovoid and stains indistinctly, whereas the cytoplasm usually stains a translucent pale green. Usually the nucleus is placed centrally and a single eccentrically situated nucleolus, staining a dull red colour with Papanicolaou, can be seen within it. There is a distinct nuclear rim of even thickness enclosing chromatin of regular, open texture.

In inflammatory conditions especially, proliferation of mesothelial cells occurs, and clusters are often seen in the smears. These may appear as irregular plaques or groups of six to twelve cells arranged in a characteristic "rosette". Sometimes these aggregates contain cells in mitosis. The morphology is relatively constant in any one effusion, although it may vary in characteristics between fluids. As with the normal cell, the limiting membrane is indistinct, but it can be seen to be roughly round and possessing no sharp angles. The staining characteristic of the cytoplasm with Papanicolaou's stain varies between translucent blue - green to a dull pink. Invariably the nuclear membrane is seen as a distinct even rim enclosing a more regular, but coarser, than normal chromatin reticulum. The nucleolus is distinct and often stains brighter than the normal. Multinucleate cells may be present, the nuclei replicating one another. In some "inflammatory" smears the cells may be "swollen", i.e., much larger, but with the same general characters except an overall reduction in staining intensity. The constancy of replication of these cells in a particular smear usually allows confident identification. Confusion may be caused initially in differentiating mesothelial cells from histiocytes, especially when an occasional cell exhibits cytoplasmic inclusions. The even texture of both cytoplasm and chromatin and the absence of the nuclear hof are helpful distinguishing features.

Malignant cells

A fundamental difference between smears from effusions containing malignant cells and malignant smears from other sites is that the neoplastic elements have arisen from a primary focus in some other organ (see Fig. 46). They are usually present in large numbers, often as intact clones, that may attempt to reproduce the architecture of the tissue from which they stem. Generally, identification will not, therefore, depend on a few cells scattered over several smears, except where the slides have been prepared from unsatisfactorily collected material. When present, the cells conform to the criteria for malignancy outlined under Abnormal Cells. Noticeable features are anisocytosis, variation in nuclear chromatin with irregular nuclear rim and large nucleoli.

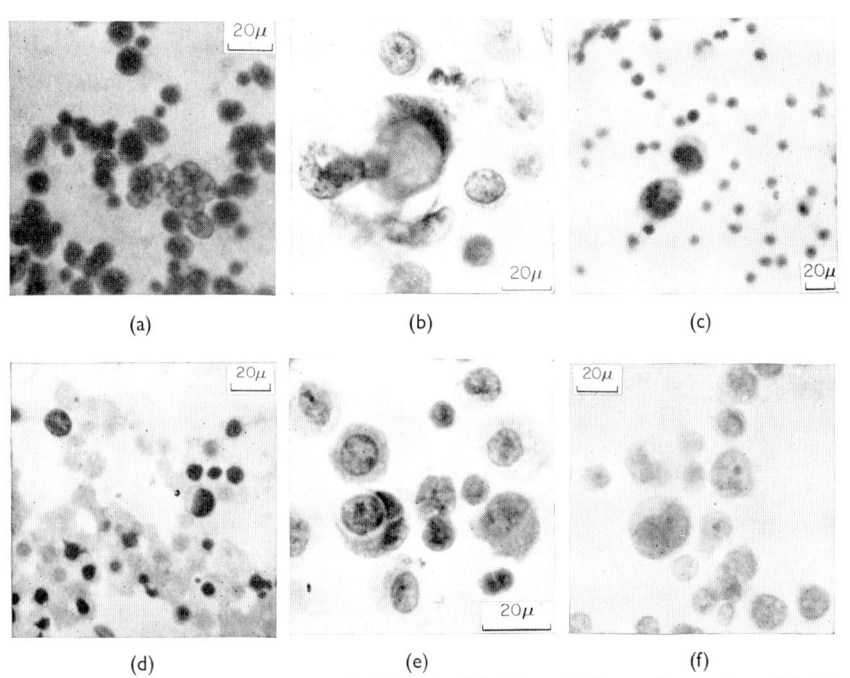

FIG. 46. Malignant cells in serous fluids: (a), (b), (c) and (d), are from pleural fluid; (e) and (f) are from peritoneal fluid.

Group II. Non-epithelial Cells

Smears may show only non-epithelial cells either of one type or a mixture in varying ratios. Attempts to assess proportions in terms of a differential count are of little value, although the predominant cell should be indicated.

All these cells retain the characteristics described under Normal Cells.

Blood

Identification of non-haematogenous malignant cells circulating in peripheral blood is not easy. Generally, the cells are ill constructed for survival in the turbulently flowing plasma. The continual buffeting causes not only the disintegration of cell clusters but also fragmentation of individual cells.

Normal Cells

In addition to the normal circulating white blood cells, occasional squamous cells may be seen. These are endothelial cells detached from the cell lining of the vein during venipuncture. Description of these cells is given under Normal Cells (pp. 37, 66).

Atypical Cells

Criteria for recognizing these cells in Papanicolaou-stained smears are:

1. The cells are generally larger than normal white blood cells.
2. The nuclei show irregular chromatin patterns with large cleared areas or hyperchromia, or both.
3. The nuclei may show folding or lobulation.
4. The nuclei may show a large nucleolus or multiple nucleoli.
5. Irregular thickening of the nuclear membrane.

In preparations stained with acridine orange (see Ancillary Technique—Fluorescence Microscopy, p. 115), the nuclei of mature white blood cells exhibit green fluorescence, although the cytoplasmic RNA shows very little red fluorescence. Neoplastic cells, rich in cytoplasmic RNA fluoresce a bright red.

Preparation of Smears

In this method (de Mello and Pimenta, 1963), 5 ml of venous blood is collected into a 10-ml disposable syringe containing 0·1 mg of heparin. Five millilitres of isotonic saline containing 200 mg of fibrinogen are drawn into the syringe and mixed. The syringe is left to stand, nozzle upwards, in a rack for 30 min at room temperature. By bending the needle into a U shape, the supernatant liquid is transferred into a centrifuge tube (see Fig. 47), which is then centrifuged at 3,500 rev/min for 10 min.

After removal of the supernatant liquor from the centrifuge tube, smears are prepared from the deposit and fixed in alcohol - ether solution.

Gastric Juice

Gastric juice collected without proper precautions may contain epithelial contaminants from the buccal mucosa, oesophagus and respiratory tract. These cells can usually be distinguished from the columnar cells of the gastric mucosa.

The Normal Smear

The columnar cells of the gastric mucosa are not exfoliated in large numbers; a properly collected specimen from a normal individual will yield smears containing only a few cells (see Figs. 48 and 49). Often most of these cells are fragmented and appear as naked nuclei. The chromatin is structureless and stains a deep lavender colour. Some of the nuclei may show considerable distortion as a result of damage during preparation. An occasional neutrophil may be present. Intact columnar cells may be dislodged by the rubber catheter used for the aspiration. These cells are about twice the diameter of the nucleus. The cytoplasm stains a pale green with a clearly seen limiting membrane. Nuclear detail shows a characteristic pattern with a fine reticulum of chromatin contained within a clearly stained, regular nuclear membrane. A single nucleolus may be present.

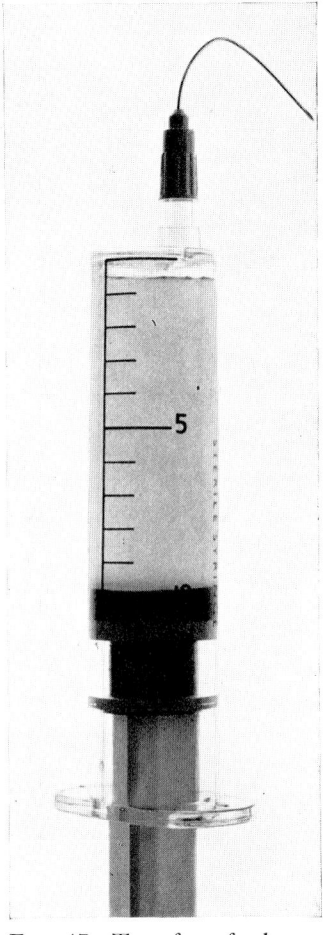

FIG. 47. Transfer of plasma in the preparation of smears from blood.

Abnormal Smears

Cells are more numerous in abnormal smears, and frequently clumps of columnar cells of regular architecture may be seen. The cells of these groups conform to the normal cells already described. Secretory vacuoles containing mucin may be seen, and when present in a free cell the nucleus is displaced and distorted, producing a typical "signet-ring" appearance.

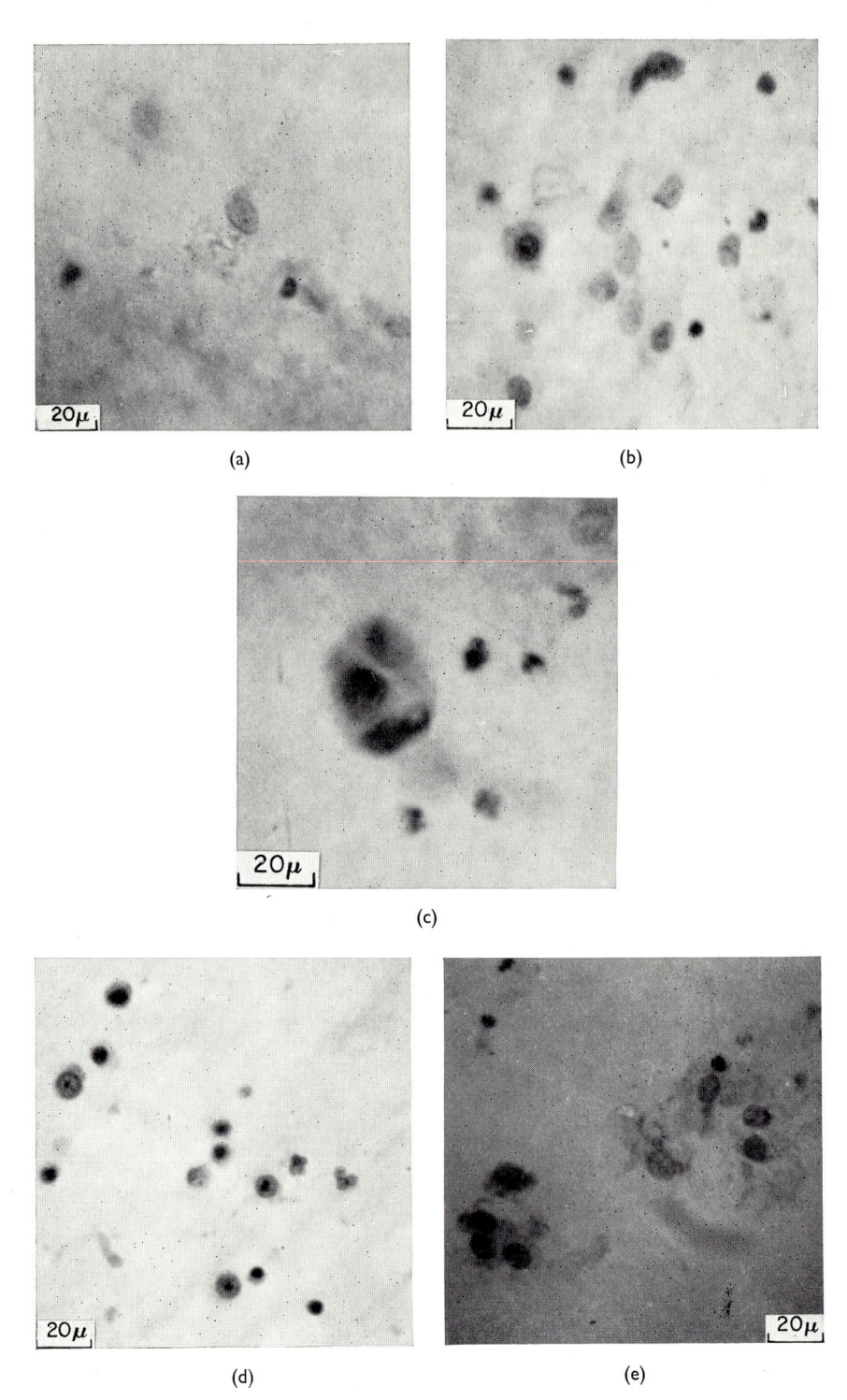

FIG. 48. Cells from gastric mucosa: (a), (b), (d) and (e) are benign; (c) is malignant.

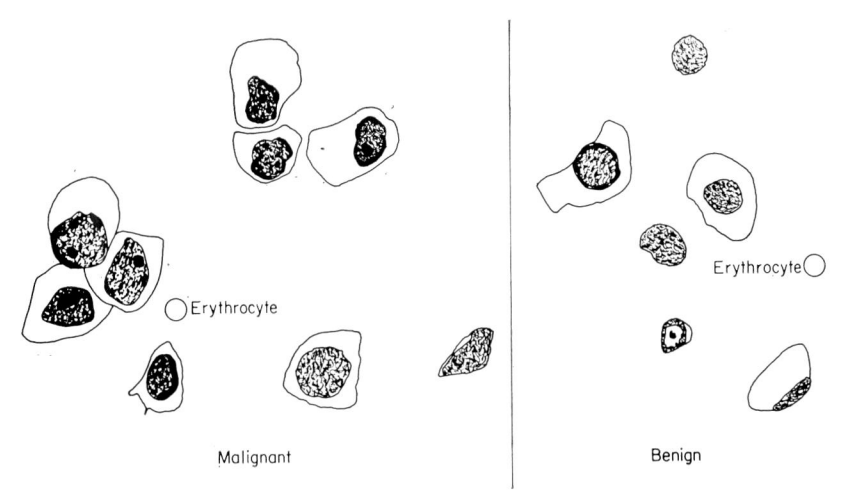

FIG. 49. Cells seen in smears of gastric aspirates.

Intracellular neutrophils in varying numbers may be present and may show evidence of degeneration.

Neutrophils, lymphocytes, plasma cells and eosinophils may be seen all staining with their usual characteristics. Erythrocytes may also be present and at times may show "ghosting" (loss of haemoglobin).

Various cell types can be seen in malignancy. Large numbers of columnar cells should arouse suspicion. Invariably these cells show alteration in the nuclear - cytoplasm ratio. The nucleus is eccentrically placed and may show hyperchromia—at times almost black—or an open, irregular chromatin pattern of coarse granules with large, irregular cleared areas. The nuclei may contain large nucleoli.

Collection of Material

Material is obtained by a rubber catheter passed into the stomach of the fasting patient, who is placed in an upright sitting position. On the morning of the test residual stomach contents are aspirated and discarded. Twenty millilitres of isotonic saline are then injected vigorously through the catheter with a syringe. Almost immediately afterwards, the saline is withdrawn and transferred to a clean 100-ml glass container. A similar procedure is followed with two further 20-ml volumes of saline, with the patient leaning first to the right and then to the left. Repeated injection and withdrawal of the same volume of saline is not advised, since this will cause disintegration of the cells. There is some risk of inducing haemorrhage during aspiration if the gastric mucosa is diseased. Preliminary irrigation with the addition of mucolytic agents (papain and α-chymotrypsin) to the lavage has been suggested.

Preparation of Smears

Pooling the three saline aspirates usually results in a specimen volume of about 50 ml. This is transferred to a 50-ml centrifuge tube and centrifuged at 1,500 rev/min for 10 min. After removal the supernatant liquor, one or two drops of 5% albumin in water are added to the deposit, and a series of smears prepared. These are fixed immediately in alcohol - ether solution and stained by Papanicolaou's method.

A critical review of gastric cytology is given by Schade (1959).

Urine

The length of the renal tubule is variously lined with simple epithelia of columnar, cuboidal and squamous types that severally contribute to the manifold functions of this structure. Transitional epithelium covers the calyces and pelves of the kidneys, the ureters and bladder. In the urethra, the epithelial surface changes to a stratified squamous type with areas of pseudo-stratified columnar cells.

Cytology

Voided normal urine contains only a few cells. Smears prepared from the centrifuge deposit of a random specimen will show a few superficial squamous cells from the urethra, together with an occasional neutrophil. Admixture by cells exfoliated from the genital tract occurs in urine from normal females. Except during menstruation these cells mainly comprise superficial squamous cells from the vagina and ectocervix, with varying numbers of neutrophils and histiocytes. Catheterization can eliminate these contaminants and, in normal individuals, this procedure yields an almost cell-free urine.

The morphology of the cells that are present rarely deviates from the normal. Urine collected by catheterization may contain cells exfoliated from the transitional epithelial surfaces. Although pleomorphic, these cells generally exhibit a roughly ovoid or squamoid shape with a large (8 to 10 μ in diameter), pale nucleus (see Figs. 50 and 51). The chromatin is of fine regular texture and the nuclear rim is of regular thickness. Some cells may show multi-nucleation (4 to 5 nuclei), the constituent nuclei replicating one another in texture and shape, but not size; occasionally a more densely staining cell with a single hyperchromatic nucleus may be seen. The staining characteristic of the cytoplasm may be eosinophilic or cyanophilic.

Inflammation, as in cystitis and pyelitis, results in exfoliation of increased numbers of transitional epithelial cells, together with large numbers of neutrophils and perhaps a few red cells. The transitional cells, which occur

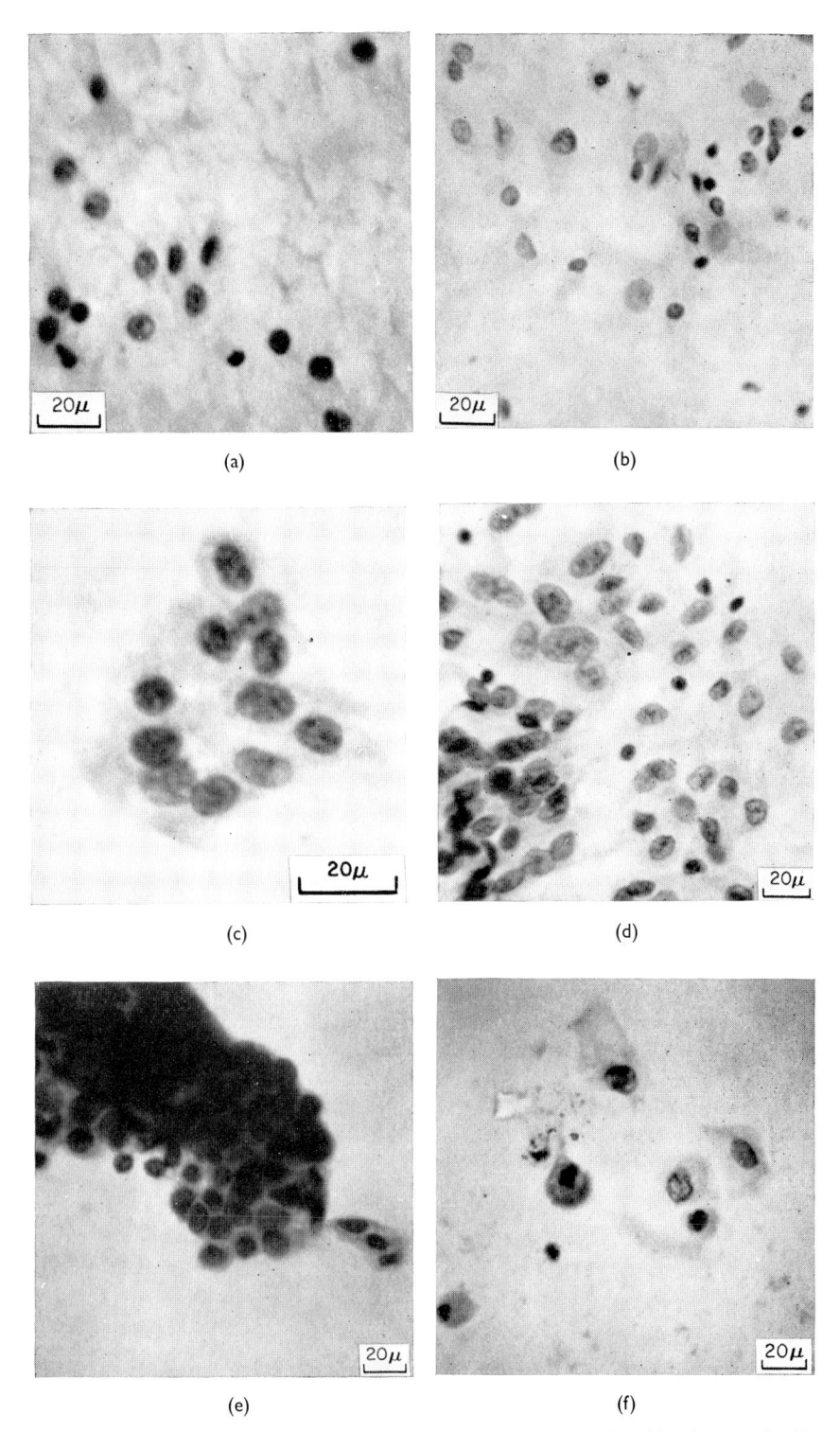

FIG. 50. (a), (b), (c) and (d) are non-malignant transitional cells; (e), cluster of cells of prostatic origin; (f), malignant cells of renal origin.

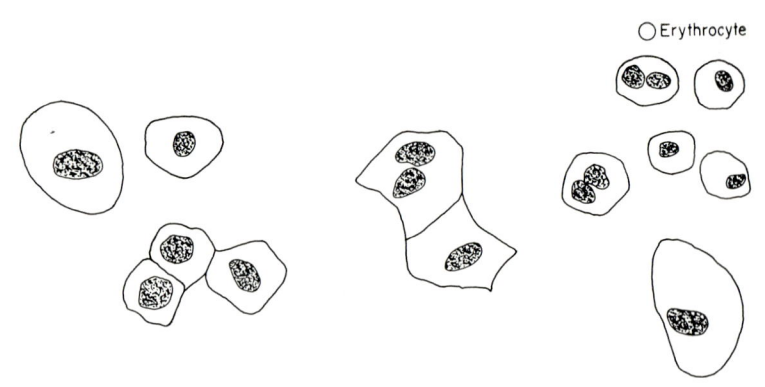

FIG. 51. Transitional cells in urine-deposit smears.

singly and in small clusters, are generally smaller than normal with hyperchromatic nuclei and indistinct cell borders. Frequently examples of the cells are seen containing single or numerous neutrophils; less often larger cells containing several hyperchromatic nuclei may be seen. The cytoplasm of the cell is eosinophilic.

In nephritis, smaller numbers of neutrophils are present. Casts may also be seen, the matrix of which is eosinophilic with distorted epithelial cells sometimes adherent to it. The cuboidal and columnar cells of the tubule epithelium are fairly easily recognized. The columnar cells are the short (cubical) form. Red blood cells are usually seen, although they may have lost their haemoglobin.

Malignant Cells

In the urine, these cells most commonly arise from a tumour in the bladder. They conform with the general criteria for malignancy. There is obvious alteration of the nuclear - cytoplasm ratio, and irregular chromatin patterns. Cell borders are indistinct and numerous atypical cells are present singly and in clusters.

Collection

About 100 ml is collected in a clean glass container. It is preferable to choose a mid-morning specimen rather than the first morning urine. This is because some of the exfoliated cells will have remained for several hours in the concentrated, overnight urine with the consequent risk of distortion and grossly modified staining characteristics.

Preparation

The specimen is gently mixed to re-suspend the sediment, and about 15 ml are removed for a routine urine analysis. This is a most helpful step and should not be omitted. A minimum routine should include tests for pH and protein on the supernatant, and a microscopic examination of a wet film of the deposit. The addition of one drop of 1% aqueous toluidine blue to this fresh preparation facilitates the identification of any cells that may be present.

The remainder of the specimen is transferred to two 50-ml centrifuge tubes, after the pH has been adjusted to about 7·0. This step is necessary to avoid alteration in the staining characteristics of the cells. The tubes are centrifuged at 1,500 rev/min for 5 min. A pipette, attached to a water-pump, is used to remove the supernatant liquor except for the last 4 to 5 ml, after which the deposits are re-suspended and pooled in a 15-ml conical centrifuge tube. Further centrifuging at 1,500 rev/min for 5 min concentrates all the cells into a single deposit. The supernatant is then removed as completely as possible without loss of cells. Two drops of 5% bovine albumen in water added to the deposit produce a cell suspension sufficient for four or five smears. If the final deposit is heavily blood stained, the upper layer, containing the nucleated cells, may be transferred to a haematocrit tube and re-centrifuged before the smears are prepared.

A technique practised by Meisels (1963) is to mix the total volume of a freshly passed urine specimen with an equal volume of 95% alcohol and to centrifuge the mixture at 2,000 rev/min for 15 to 20 min. The deposit is smeared and fixed in alcohol - ether containing 1% celloidin.

In the absence of a visible deposit, some workers pass the urine through a Millipore screen, which is then processed and stained (see Ancillary Techniques—Filtration, p. 122). Umiker (1964) advocates the use of acridine orange for staining smears from urine (see Ancillary Techniques—Fluorescence Microscopy, p. 115).

Other Fluids and Secretions

Only a small number of breast and prostatic secretions and cerebrospinal fluid (CSF) specimens have been examined in this laboratory.

Breast Secretion

Smears are prepared by collecting exudate from the nipple directly on to slides, which are processed by the alcohol - ether - Papanicolaou technique (Kjellgren, 1964).

Most exudates yield smears containing only a few cells that may include keratinized squamous cells from the nipple, red cells, neutrophils and histiocytes. Exudates in mastitis contain numerous examples of these cells.

Many of the histiocytes possess a nucleus with a finely granular chromatin pattern and containing one or two nucleoli: the cytoplasm is often foamy.

In malignancy, small plaques of secretory cells from the ducts may be seen.

Prostatic Secretion

Smears are made by collecting the secretion, obtained by prostatic massage, directly on to glass slides. These are processed by the alcohol - ether - Papanicolaou technique (Koss and Durfee, 1961).

The origin of some of the epithelial cells seen is often difficult to decide. A typical smear representative of the small series examined in this Unit would contain an occasional squamous cell (probably from the urethra), spermatozoa and an occasional cluster of small cells (see Fig. 50). These cells possess regular, condensed, blue - black nuclei eccentrically placed in an almost colourless cytoplasm. A few neutrophils and histiocytes with an occasional red cell are usually present. *Corpora amylacea* (calcified prostatic secretion) may also be seen.

CSF

The total volume of fluid received is centrifuged at 2,000 rev/min for 5 min (Naylor, 1964). As much of the supernatant liquor as possible is removed by inverting the tube, which is then allowed to drain. Normal fluids will contain only a few lymphocytes.

CHAPTER VII

Documentation

Before a patient's slides and report are integrated into the permanent file of a cytology unit, it is essential to have comprehensive, accurate data. A request form carrying adequate information to identify the patient, together with essential clinical data should accompany every specimen. The request forms used in this Unit are illustrated in Fig. 52.

Each consists of three copies interleaved with carbon paper. This arrangement allows the worker examining the slides to make his own report without the intervention of a typist. At the same time the report is entered in an abbreviated form on a previously prepared file card, which need be no larger than 5 ×3 in. As the index of cards accumulates, the time required to search the file increases. A space- and time-saving apparatus, the Kardveyor (Remington-Rand Ltd) (see Fig. 53), is available in a range of capacities extending to 500,000 cards.

The information on the specimen bottle should be carefully checked against that given on the request form and any errors and omissions resolved before the slides are processed. If anomolous data are documented, then subsequent review of a patient's slide, perhaps 2 or 3 years later, may render the re-examination valueless.

Every slide examined and a copy of every report issued should be catalogued. A large proportion of cytology specimens, for example gynaecological material, will be from patients on a register for periodic review, and any change in the slide-grading over the intervening period will be of significance. If a biopsy is subsequently performed, the histology reference number should be recorded on the file card. Ideally, a histological section illustrating the report should be filed in the wallet containing the smears. After the request has been entered on a day register, the file is searched to determine whether the patient is already on record. Record cards are filed alphabetically and, in addition, a carbon copy of each report issued is filed in numerical sequence. The use of composite request forms with interleaved carbons means that the filed back copy is an exact duplicate of the actual report issued.

Coding Information

Statistical analysis can be greatly facilitated if the report is coded on the file card. A punched-card system is used in this department, which allowed

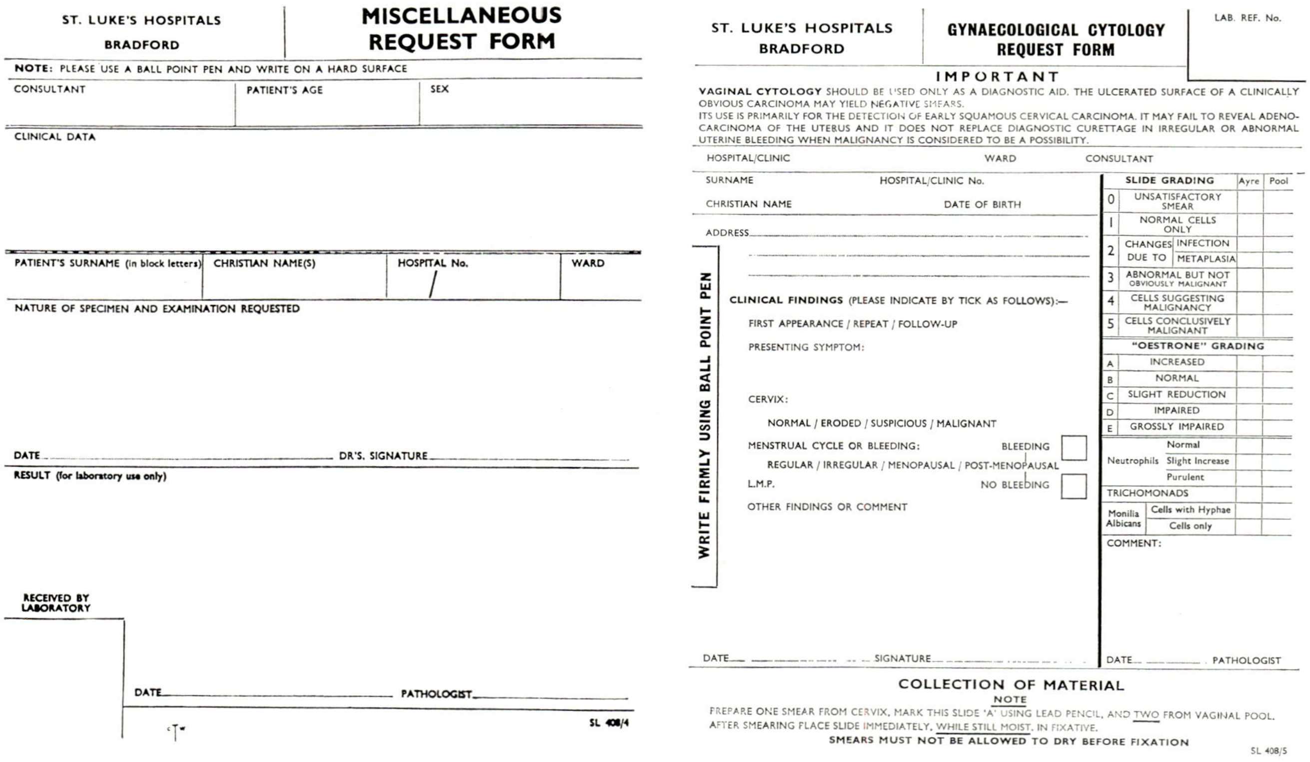

ST. LUKE'S HOSPITALS
BRADFORD

MISCELLANEOUS
REQUEST FORM

NOTE: PLEASE USE A BALL POINT PEN AND WRITE ON A HARD SURFACE

CONSULTANT	PATIENT'S AGE	SEX

CLINICAL DATA

PATIENT'S SURNAME (in block letters)	CHRISTIAN NAME(S)	HOSPITAL No.	WARD

NATURE OF SPECIMEN AND EXAMINATION REQUESTED

DATE ____ DR'S. SIGNATURE ____

RESULT (for laboratory use only)

RECEIVED BY LABORATORY

DATE ____ PATHOLOGIST ____

SL 408/4

ST. LUKE'S HOSPITALS
BRADFORD

GYNAECOLOGICAL CYTOLOGY
REQUEST FORM

LAB. REF. No.

IMPORTANT

VAGINAL CYTOLOGY SHOULD BE USED ONLY AS A DIAGNOSTIC AID. THE ULCERATED SURFACE OF A CLINICALLY OBVIOUS CARCINOMA MAY YIELD NEGATIVE SMEARS.
ITS USE IS PRIMARILY FOR THE DETECTION OF EARLY SQUAMOUS CERVICAL CARCINOMA. IT MAY FAIL TO REVEAL ADENO-CARCINOMA OF THE UTERUS AND IT DOES NOT REPLACE DIAGNOSTIC CURETTAGE IN IRREGULAR OR ABNORMAL UTERINE BLEEDING WHEN MALIGNANCY IS CONSIDERED TO BE A POSSIBILITY.

HOSPITAL/CLINIC WARD CONSULTANT

SURNAME HOSPITAL/CLINIC No.
CHRISTIAN NAME DATE OF BIRTH
ADDRESS ____

WRITE FIRMLY USING BALL POINT PEN

CLINICAL FINDINGS (PLEASE INDICATE BY TICK AS FOLLOWS):—
FIRST APPEARANCE / REPEAT / FOLLOW-UP
PRESENTING SYMPTOM:
CERVIX:
NORMAL / ERODED / SUSPICIOUS / MALIGNANT
MENSTRUAL CYCLE OR BLEEDING: BLEEDING
REGULAR / IRREGULAR / MENOPAUSAL / POST-MENOPAUSAL
L.M.P. NO BLEEDING
OTHER FINDINGS OR COMMENT

	SLIDE GRADING		Ayre	Pool
0	UNSATISFACTORY SMEAR			
1	NORMAL CELLS ONLY			
2	CHANGES DUE TO	INFECTION		
		METAPLASIA		
3	ABNORMAL BUT NOT OBVIOUSLY MALIGNANT			
4	CELLS SUGGESTING MALIGNANCY			
5	CELLS CONCLUSIVELY MALIGNANT			
	"OESTRONE" GRADING			
A	INCREASED			
B	NORMAL			
C	SLIGHT REDUCTION			
D	IMPAIRED			
E	GROSSLY IMPAIRED			
Neutrophils	Normal			
	Slight Increase			
	Purulent			
TRICHOMONADS				
Monilia Albicans	Cells with Hyphae			
	Cells only			

COMMENT:

DATE ____ SIGNATURE ____

DATE ____ PATHOLOGIST

COLLECTION OF MATERIAL

NOTE

PREPARE ONE SMEAR FROM CERVIX, MARK THIS SLIDE 'A' USING LEAD PENCIL, AND TWO FROM VAGINAL POOL.
AFTER SMEARING PLACE SLIDE IMMEDIATELY, WHILE STILL MOIST, IN FIXATIVE.
SMEARS MUST NOT BE ALLOWED TO DRY BEFORE FIXATION

SL 408/5

FIG. 52. Cytology request forms.

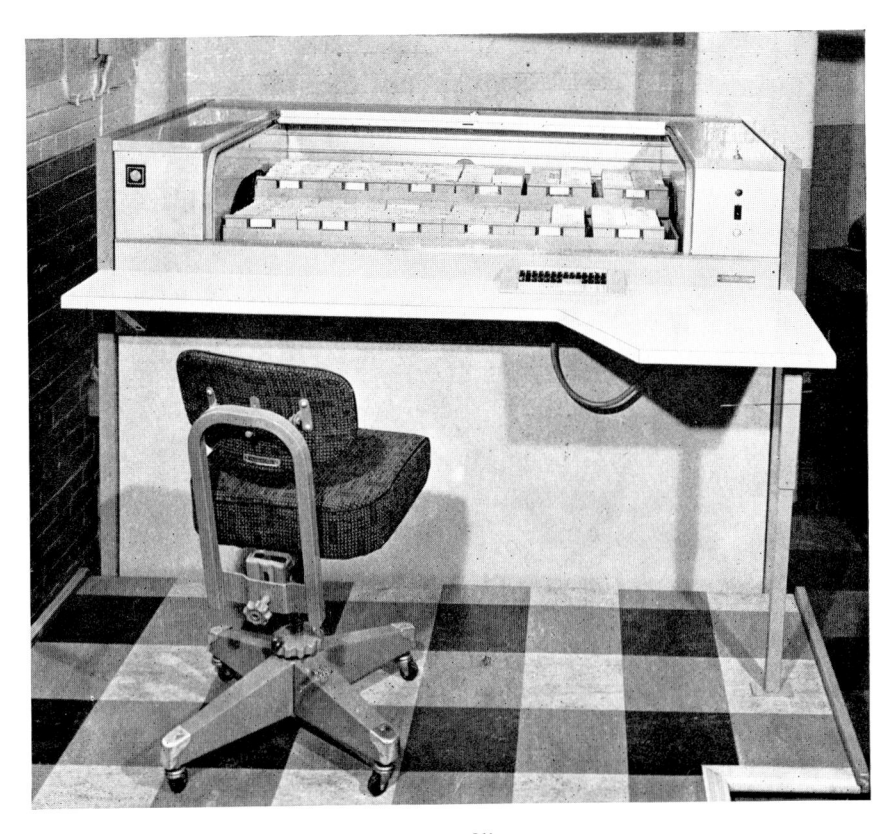

FIG. 53. Kardveyor filing apparatus.

the continued use of existing cards, since about 1,400 cases were already on record. Holes were drilled into the cards and the code illustrated was devised (see Fig. 54). Reports can be translated with an ordinary card punch. This method has obvious limitations in respect of the amount of information that can be coded. Information concerning directional changes in slide or hormone grading is not automatically obtained. It does, however, reduce the number of cards that have to be handled in order to extract this information.

Information is obtained by passing sorting needles through blocks of the cards along the holes required (see Fig. 55). If, for example, three needles are used, the first through "Papanicolaou grade 3", the second through "Oestrone grade B" the third through "Trichomonads", when the block is inverted only those cases that have been graded 3, oestrone B, and trichomonas "positive" will fall from the block of cards.

108 PRACTICAL CYTOLOGY

Papanicolaou grading Oestrone grading Blood Normal numbers } Neutrophils Increased numbers Trichomonads Monilia Yeast cells
0 1 2 3 4 5 A B C D E

Papanicolaou grade 2
Oestrone grade C
Increased neutrophils
Trichomonads present

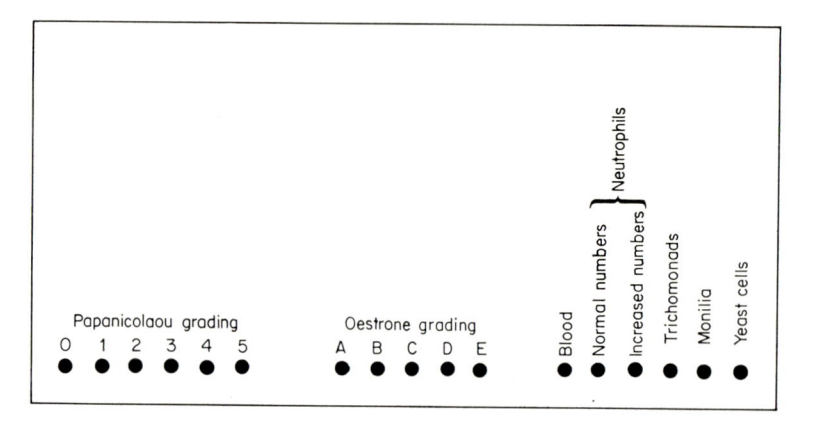

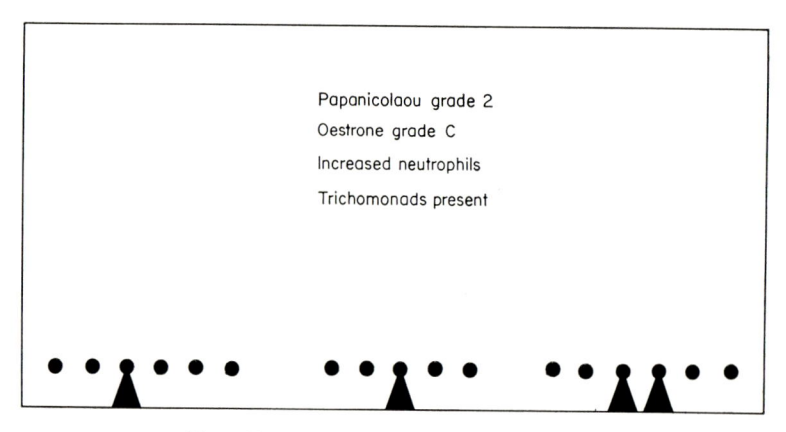

Fig. 54. A simple punched-card system.

Surveillance Chart

In this laboratory, all gynaecological patients whose smears produce results suggesting suspicion or malignancy (grade 3 and above) and all histologically diagnosed uterine cancers are recorded on a broadsheet some 3 ×2 ft (McInroy, 1966). This is conveniently placed on a "drop-leaf" board in the records office (see Figs. 56 and 57), and carries the grading assessment of each person who has examined the slides.

One purpose of the chart is to ensure that a smear reported as positive or suspicious is followed-up; this is most important when suspicion is aroused in an apparently well woman. The chart is also a useful guide to the slide-grading accuracy of individual members of the team.

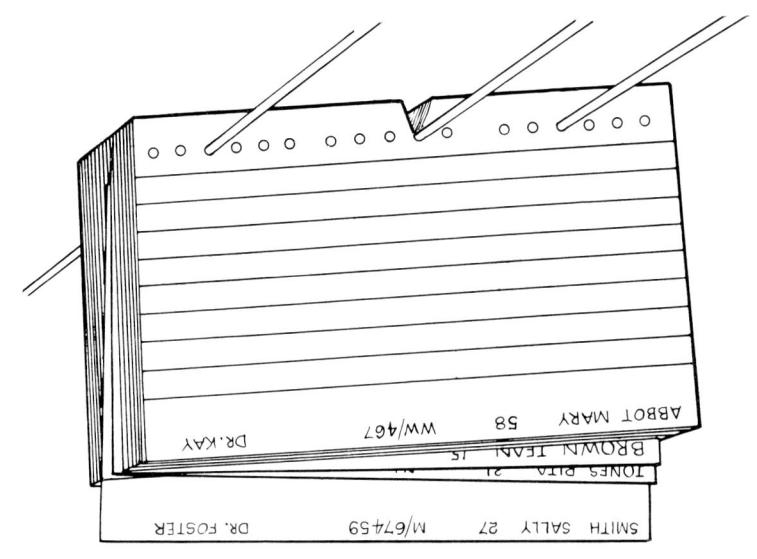

FIG. 55. Retrieving coded data.

Storage of Slides

One of the problems in a busy cytology unit is the storage of a large accumulation of slides. Patients on regular review will remain on the register for long periods. Apparently well women, for example, will have smears repeated at intervals of 3 years or so. Frequently, it is desirable to re-examine the previous slides from such cases. These latter must, therefore, be readily to hand; it is impracticable to store slides and reports of previous years in a remote basement. Ideally a room adjacent to the office should be set aside for these files. In this Unit, slides are stored in steel cabinets that contain 44 drawers. Each drawer holds 600 slides, mounted in cardboard wallets and filed in numerical sequence. The capacity of the drawers may be increased to 800 by using a slotted Perspex or wood base in the drawer instead of the wallets.

Specially constructed consoles for filing slides are available (Autotechnicon Ltd), but these are relatively expensive in terms of cost and space.

Preparation of a Reference Collection

A considerable saving in the time required to train initiates can be achieved by accumulating an index of smears that contain classified cells. During the routine examination of smears examples of cells will be seen that can be identified without hesitation. The location of such cells on the slide can be marked, by using the ink-dot technique previously described. The slide number can be catalogued and, by using a card index, a collection

110 PRACTICAL CYTOLOGY

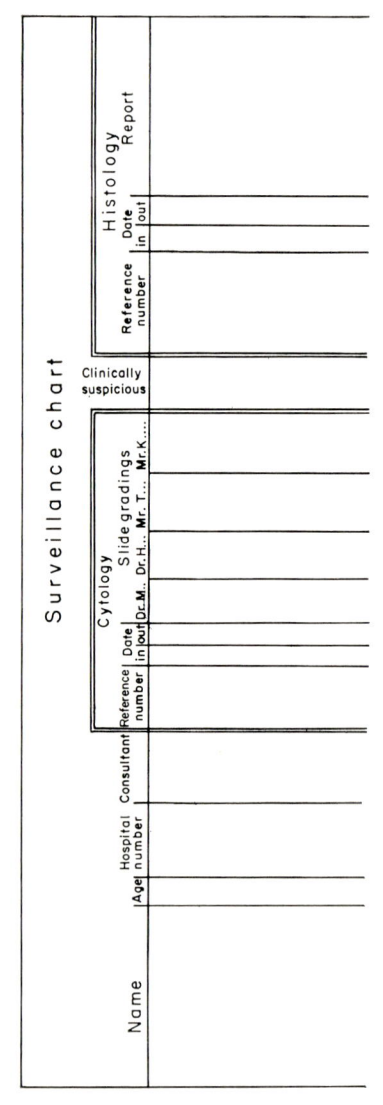

FIG. 56. Surveillance chart.

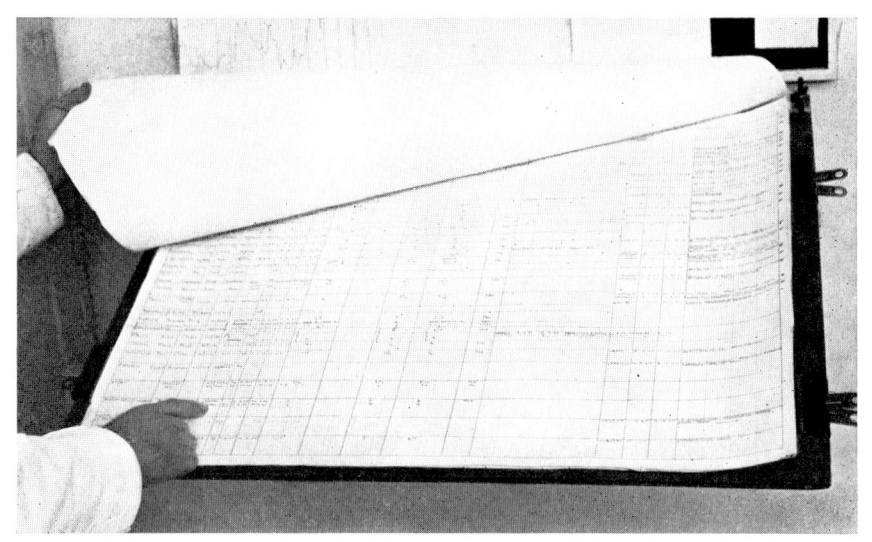

FIG. 57. "Drop leaf" board for surveillance chart.

can be built up from these slides. Other cells, which cause difficulty initially, are often identified when a critical study of the subsequent histological section is carried out. These reference cells form an invaluable part of the collection, both for personal reference and for students. It is wise in these instances to include a duplicate of the histological section in the cytology-slide wallet.

CHAPTER VIII

Ancillary Techniques

Fluorescence Microscopy

A body is said to fluoresce if it re-emits light of a longer wavelength than that which is incident upon it and if the emission ceases when the incident light is removed. The fluorescence, therefore, will be of a different colour than the incident, exciting light, the change being to a colour toward the red end of the spectrum.

When ultraviolet radiation, which is not visible, is the exciting agent, a fluorescing body will usually re-emit this as light within the visible spectrum. The wavelength of the emission (and, therefore, the colour), is a function of the molecular density of the body. A structure fluorescing red light has absorbed more energy from the incident radiation than one that fluoresces a green or yellow colour.

In fluorescence microscopy, a light source rich in violet - blue radiation (400 to 500 mμ wavelength) is chosen. Sequences have been devised to attach fluorescing agents (fluorochromes) to cell structures. A smear is first "stained" with a fluorochrome solution and then differentiated. When such a preparation is excited with this light, material without the fluorochrome attached remains invisible, whereas cell components with the fluorochrome appear in a range of colours.

Problems associated with this technique are the light source, the filtration of the light produced and autofluorescence.

The Light Source

The energy emissions required to excite the fluorochromes used lie in the shortest wavelengths (violet - blue) of the visible spectrum (400 to 500 mμ). Three methods of obtaining this energy are available.

(a) *Mercury-vapour lamp*

A small amount of mercury is contained within the glass envelope together with two heavy tungsten electrodes, which have a heating element wound around them. When the lamp is switched on, the elements heat the tungsten to incandescence, the mercury vaporizes and there is a discharge across the electrodes, producing an emission rich in ultraviolet and possessing some visible light. The intensity of the emission can vary markedly at times, owing to mercury condensing on the electrodes and so cooling them. A great deal of unwanted low ultraviolet and infrared light (and therefore heat) is generated by the apparatus, which has a short life and is expensive.

(b) *Tungsten-filament lamp*

A conventional tungsten-filament, high-intensity lamp of the type used in cinematograph projectors can be used. It is necessary to "over run" the lamp—to use a voltage higher than that for which the lamp was designed in order to obtain a sufficiently intense emission by raising the working temperature of the filament. This also produces a considerable amount of heat. Over running greatly shortens the life of the lamp, owing to evaporation of the tungsten filament, which condenses on the glass envelope as a black film.

(c) *Tungsten iodine lamp*

This apparatus is the best choice for routine fluorescence work. It is much cheaper than the mercury-vapour lamp and can be slightly "under run" to give a working life in excess of 100 h. Much less heat is generated, the lamp is more durable and requires no starting chokes. The iodine present in the quartz envelope combines with the tungsten vapour, thus preventing blackening of the envelope. Near the filament the high temperature causes the tungsten iodide to dissociate, with the consequent re-deposition of tungsten. The lamp is a standard fitting on the Gillett and Sibert multi-purpose "Conference" microscope.

Filters

Two filters are required: one is interposed between the light source and the condenser (see Fig. 58). This is called the primary filter and usually has a band pass of 360 mμ to 460mμ. Light of these wavelengths passes through the condenser, which brings the rays to a focus at the surface of the slide carrying the object. If the object does not autofluoresce or possess an attached fluorochrome, the rays pass through the object, which remains non-luminous. The rays, rich in selected radiation, pass through the optics of the microscope with diminishing intensity, owing to absorption by the lenses, until they are stopped by a second filter. The secondary filter is necessary to stop light transmission below a chosen level.

Autofluorescence

Objects that autofluoresce are said to possess primary fluorescence: they absorb the incident short-wavelength radiation, re-emitting it as visible light (often in the yellow - green region) and so appear coloured.

Material that retains the fluorochrome is said to possess secondary fluorescence, re-emitting the incident radiation in the visible spectrum. The colour and intensity of the re-emission is determined by the molecular structure of both the fluorochrome and the substance to which it is attached, as well as the nature of the incident radiation.

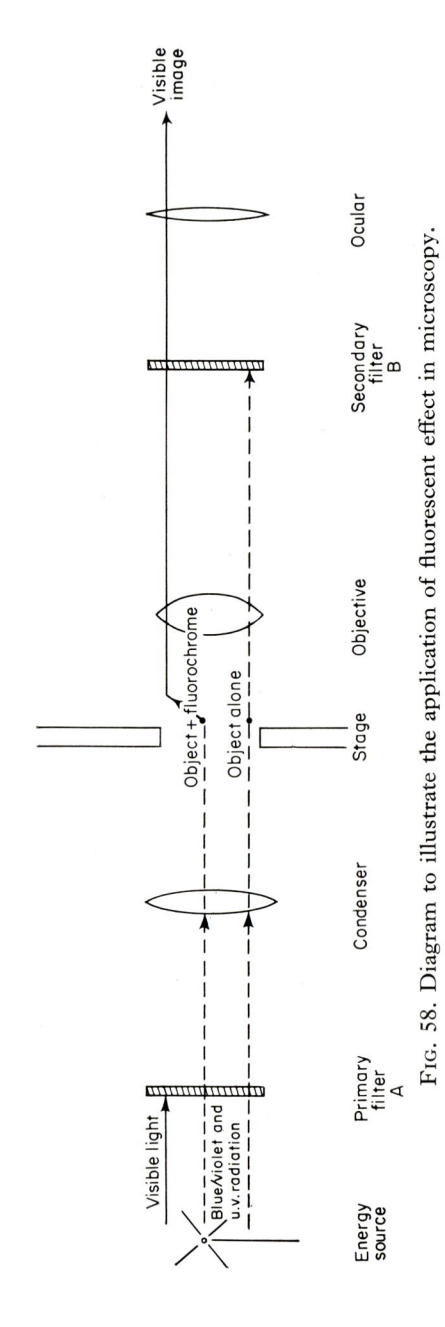

FIG. 58. Diagram to illustrate the application of fluorescent effect in microscopy.

In medical cytology, the fluorochrome commonly used is acridine orange. The background of the microscope field is black, against which the "stained" (fluorescing) structures stand out in brilliant colours. Maximum resolution is obtained by excluding all extraneous light, either by working in a darkened room or by using a microscope hood. It is not necessary to use special lens systems for this technique; if immersion oil is used, a non-fluorescing preparation is required, e.g., Fluorfree (G. T. Gurr Ltd).

Acridine Orange Staining Technique

In this method (Bertalanaffy, 1960) alcohol - ether fixed smears should be used.

1. Alcohol, 80%   10 sec.
2. Alcohol, 70%   10 sec.
3. Alcohol, 50%   10 sec.
4. Distilled water   10 sec.
5. Acetic acid, 1%   10 sec.
6. Distilled water   10 sec.
7. Acridine orange   3 min.
8. Buffer solution, pH 6·0   1 min.
9. Calcium chloride solution   1–2 min.
10. Buffer solution, pH 6·0   Rinse.
11. Mount in buffer

(a) *Acridine orange*

An aqueous stock solution 0·1% is diluted to 0·01% with the phosphate buffer.

(b) *Phosphate buffer, pH 6·0*

M/15 potassium dihydrogen phosphate, 2 volumes, plus M/15 disodium hydrogen phosphate, 5 volumes: see p. 137 for instructions for preparing the M/15 solutions.

(c) *Calcium chloride*

0·1 M (11·1 g in 1 litre of water). Wet preparations may be de-stained by immersion in 50% alcohol and subsequently stained by other techniques.

Interpretation

With correct times for staining and differentiation, acridine orange produces a red fluorescence with cytoplasmic and nuclear RNA; DNA will fluoresce green. Neoplastic cells possess high concentrations of RNA and, therefore, fluoresce an intense red.

Rapid Wax Embedding

Cell deposits, clots and fragments of tissue when present in cytological specimens often yield more information if processed by histological techniques rather than by smearing the material on slides. In sections, cell clusters frequently reveal evidence of the architecture of the parent tissue.

Preparation

Any deposit in the fluid is re-suspended by gentle mixing. Fragments of tissue quickly sediment on standing, thus allowing withdrawal of the supernatant suspension of cells by pipette. This suspension is processed by cytological techniques (see Concentration Methods, p. 121). The fragments are re-suspended in 5 ml of fixative. Clots, if present, can be picked out and placed directly into fixative. Centrifuge deposits may be treated by replacing the supernatant fluid with about 5 ml of fixative.

Fixation

The material is processed in a 15-ml round-bottomed centrifuge tube, stoppered with a bark cork. Two types of fixatives may be used: the cytological fixatives of Carnoy and Papanicolaou; and the cytomicroanatomical solutions of Zenker and Heidenhain ("Susa"). Both Carnoy's and Papanicolaou's solutions, which accomplish fixation in 20 min, cause considerable shrinkage. Zenker and "Susa" solutions produce better fixation, but require 1 h to accomplish this. Since both reagents contain mercuric chloride, sections must be treated with iodine - thiosulphate before staining. "Susa" solution is recommended.

Dehydration

The tube is briefly centrifuged, and the supernatant fixative discarded. About 10 ml of acetone is added, the stopper replaced and the contents mixed by careful inversion. This acetone is then immediately removed after centrifuging, replaced with 5 ml of 74 O.P. industrial spirit, and the tube left to stand for 15 min after gentle agitation of the deposit. Three further changes of alcohol are carried out at 15-min intervals by means of a similar technique.

Clearing

After the fourth alcohol is removed, 10 ml of benzene is added, the tube re-stoppered and the contents mixed by inversion. The tube is immediately centrifuged and the benzene discarded. Five millilitres of fresh benzene are added, and the tube left to stand for 15 min after gentle agitation of the deposit. Two further changes of benzene are made in similar fashion. During the final benzene wash, the tube is placed in the embedding oven.

Wax Impregnation

Without removing the tube from the oven, the stopper is withdrawn, and all but the last drop of benzene removed by pipette. About 2 ml of wax is added and the unstoppered tube left for 5 min after gentle agitation of the deposit. By using a pipette at the temperature of the wax, without removing the tube from the oven, the wax is withdrawn, being replaced with a fresh 2-ml volume of wax. The tube is left for 10 min. Two more similar changes are made.

Blocking

The wax is withdrawn by pipette and replaced with 3 ml fresh wax (see Fig. 59). A wood applicator, its end previously impregnated with wax, is now passed down the tube so that its tip is about halfway down the column

FIG. 59. Withdrawal of wax block from centrifuge tube.

of wax. Cotton-wool wound round the upper third of the applicator, holds it in place. Applicators are stored in a 2-oz honey pot containing a 2-cm layer of wax and placed in the embedding oven.

The tube is then taken from the oven and allowed to harden at room temperature. When the wax has solidified it is cooled further by placing the tube in the freezing compartment of the refrigerator for about 10 min. To withdraw the wax block, the applicator is twisted slightly while the tube is held steady. The tip of the tube is immersed in hot water. When the glass has heated, a sudden "give" is felt in the applicator, and this is the signal to withdraw the block. A steady pull on the applicator brings out the wax cleanly without loss of material. Make sure that the tube is immersed just to the level of the wax. An additional sign can be observed as a front of molten wax moves up from the tip of the tube. Allow the surface of the block to cool before extracting the wood applicator with a sharp twist just above the surface of the wax.

The block is then squared off by paring, mounted and sectioned. No facing down should be attempted. The resulting serial sections may be mounted in 5-cm long ribbons, two or three side by side on each slide, in the usual way.

Fixatives

(a) *Cytological*

   (i) *Carnoy's Solution*

   74 O.P. spirit, 60 ml.
   Chloroform, 30 ml.
   Acetic acid, glacial, 10 ml.
   Fixation time, 20 min.

   (ii) *Papanicolaou's solution*

   74 O.P. spirit, 1 volume.
   Diethyl ether, 1 volume.
   Fixation time, 20 min.

(b) *Cyto-microanatomical*

   (i) *Zenker's solution*

   Mercuric chloride, 5 g.
   Potassium dichromate, 2·5 g.
   Sodium sulphate, 1·0 g.
   Distilled water, 100 ml.
   Before use, add:
   Acetic acid, glacial, 5·0 ml.
   Fixation time 1 h.

(*ii*) *"Susa" solution*

Mercuric chloride, 4·5 g.
Sodium chloride, 0·5 g.
Trichloracetic acid, 2·0 g.
Acetic acid, glacial, 4·0 ml.
Formalin, 20 ml.
Distilled water, 100 ml.
Fixation time, 1 h.

Removal of mercuric salts from slides

Mix 1 g of iodine and 2 g of potassium iodide in a boiling-tube and add 2 ml of distilled water, heat to boiling, cool and dilute to 100 ml.

Before staining, immerse the slides for 5 min first in the iodine - potassium iodide solution, followed by a similar immersion in 5% aqueous sodium thiosulphate. Wash the slides in gently running water.

Serial Sections

The first few sections of a biopsy may provide an explanation for atypical cells seen in the pre-operative smears. In some instances, however, serial sectioning may be considered, e.g., if the cause is thought to be at a single focus in the tissue. Casual "facing down" of these blocks may result in the discarding of important sections. It is better, therefore, to retain all of these ribbons in sequence until the stained sections have been examined. This can be accomplished by inter-leaving each ribbon between thin typing paper or polythene film and sandwiching the whole between two cardboard strips secured by rubber bands or Sellotape (see Fig. 60). These are stored

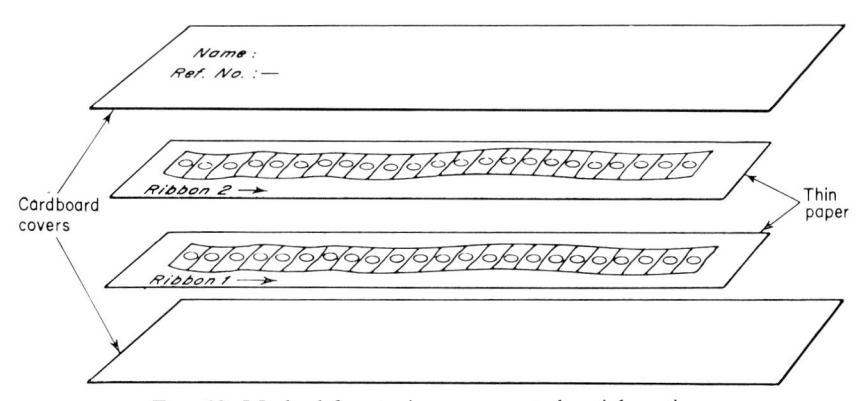

Fig. 60. Method for storing unmounted serial sections.

at 5°C and, if subsequently required, they can be mounted on slides even after a lapse of several weeks.

Before packing each ribbon, the final section is cut off and mounted. Sections representing a regular series of steps through the tissue can then be mounted in sequence on a single slide (see Fig. 61). If, subsequently, it is necessary to mount the whole of the serial sections, the large number of slides involved presents a problem in storage. This may be overcome by mounting the sections on Cellophane strips previously prepared with egg-albumen adhesive.

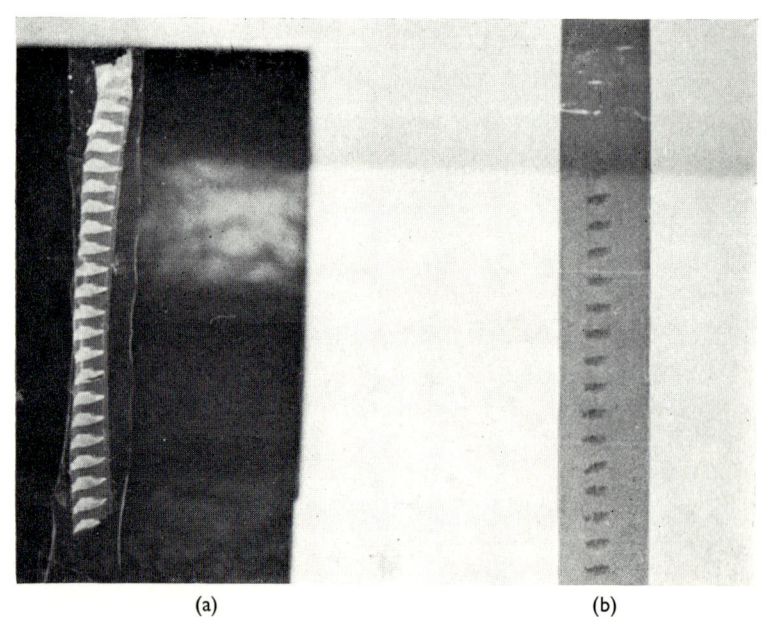

Fig. 61. Serial sections on Cellophane: (a), before staining; (b), after staining.

Processing and staining is then the same as for orthodox sections as far as the final xylene bath. The stained strips are then transferred to a bath of 50% DPX in xylene and allowed to drain and dry in air at room temperature (about 2 h). They can then be safely handled and stored indefinitely in paper envelopes.

Individual sections can be cut from the ribbons with scissors and mounted on slides in the usual way. The cut end of the ribbon should be re-sealed with DPX to exclude air.

A whole biopsy can be stored in a paper envelope, in contrast to the 200 or so slides that are needed for conventional preparations.

Concentration Methods

Many of the non-gynaecological specimens received for cytological examination are specimens of body fluids, e.g., pleural and ascitic aspirates. It is pertinent to remark here that the total volume of such fluids should be submitted for investigation in the container originally used for the collection. The practice of sending part of the specimen or of transferring it into laboratory specimen bottles involves the risk of loss of cells by sedimentation.

In the laboratory an attempt should be made at selective concentration of all the cells in the fluid. Clots and fragments of tissue that may be present should be picked out and processed separately This may be done either by preparing smears, which is likely to distort the cells or by the paraffin-wax-section technique For the remaining fluid, the ideal technique is that which will achieve recovery, without damage, of all the contained cells, separated into groups according to type, as simply as possible.

Density-gradient Methods

Neoplastic cells have been shown to have densities less than normal body cells (Seal, 1956), as is shown below (Sharpe, 1963):

Neoplastic cells	White blood cells	Erythrocytes
1·056	1·065	1·092–1·097

Simple Centrifuging

This property of the cells is utilized in the simple centrifuge technique. When a suspension of cells is centrifuged, a steady state is reached, where the cells have settled in strata according to their densities. High speeds are not necessary to accomplish this: the cells will separate satisfactorily after 15 min at 1,500 rev/min in a standard centrifuge. Speeds and times in excess of this merely cause tight packing of the deposit and consequent distortion of the cells. In addition, the higher temperatures to which the cells are exposed often alters staining characteristics. A suggested technique is to stand the specimen in the refrigerator to allow sedimentation to take place, e.g., overnight. Cooling eliminates convection currents that otherwise interfere with sedimentation of the lightest cells. On removing the container from the refrigerator, the supernatant fluid is withdrawn by a glass delivery tube attached to a vacuum pump, and the sediment re-suspended in the small volume of residual supernatant liquor. The resulting suspension is transferred to a standard 15-ml tube and centrifuged. Small deposits are more easily handled in a Wintrobe haematocrit tube. Subsequently, a numbered series of slides can be prepared at descending levels through the buffy layer with a fine-bore Pasteur pipette. The method eliminates all but the lightest erythrocytes from the smears.

Flotation

This technique adds a refinement to the simple centrifuge method, but is based on the same fundamental premise. Provided that all the cells of a type have the same known density, it follows that they and any lighter cells can be separated from more dense cells by interposing a non-miscible solution of suitable specific gravity.

Of the various reagents investigated for use in flotation techniques, silicone fluids have been shown to be the most useful. These are a group of substances whose molecules consist of straight chains of alternating silicone and oxygen atoms: the remaining silicon valencies are satisfied with various organic groups. Suitable fluids for the technique are silicones MS555 and MS710 (preparations by Midland Silicones Ltd and obtainable through Hopkin & Williams Ltd).

(a) *Solutions*

Mix 100 ml of MS555 and 25 ml of MS710; the mixture yields a fluid with specific gravity 1·075.

Dissolve 350 g of poly(vinyl pyrrolidone) (PVP) in 8 litres of isotonic saline. Add 10 ml of Triton WR1339 (Winthrop Laboratories) and 25 g of EDTA (disodium ethylenediaminetetra-acetate dihydrate) and make up to 10 litres with saline.

(b) *Method*

In this method (Sharpe, 1963), 10 ml of silicone fluid is placed in a 50-ml polypropylene centrifuge tube (MSE Ltd). Twenty millilitres of PVP solution are then carefully superimposed, followed by 10 ml of the fluid to be examined.

Without shaking, the tube is now centrifuged at 1,500 rev/min for 15 min. The supernatant liquor is removed to within 1 cm of the silicone interface, and the residual 1-cm column transferred to a second centrifuge tube by successive washing and aspiration. The resulting suspension is centrifuged at 800 rev/min for 30 min in order to concentrate the cells. Smears may be prepared from this deposit in the usual way, or the deposit may be re-suspended in formal saline before filtration.

Filtration

In this method, developed by Seal (1956), fluids containing cells are filtered through a membrane designed to retain cells while passing the fluid and small particles. Subsequently, the membrane, with adherent cells, is fixed, stained and mounted on a microscope slide.

Its chief advantage is to minimize loss of cells during concentration. Disadvantages include thick preparations, debris obscuring cells and cumbersome mounted preparation.

(a) *Cellulose membrane*

The filters are made from cellulose esters and constructed so that the pore size is known. Products by Courtaulds Ltd (obtainable from Oxoid Ltd) and Millipore (obtainable from V. A. Howe and Co. Ltd) of pore size 5 are suitable.

(b) *Apparatus*

(*i*) *Hemming filter*. This apparatus (see Fig. 62) consists of a short metal cylinder with its internal wall threaded to take a standard bijou bottle at each end. Communication between the two bottles is interrupted by a perforated diaphragm that forms the support for the filter membrane. The apparatus is obtainable from Beaumaris Instrument Co. Ltd.

With a membrane in position, the collar is screwed on to a bijou bottle containing the fluid. A receiving bottle is screwed into the other end and the apparatus centrifuged at 3,000 rev/min for 15 min. This filter is useful if the volume of fluid is less than 4 ml.

(*ii*) *Pressure filter*. This metal filter (see Fig. 63) is dismantled by unscrewing the cup revealing a carbon diaphragm with rubber ring seals on either side. The diaphragm is the support for the filter membrane.

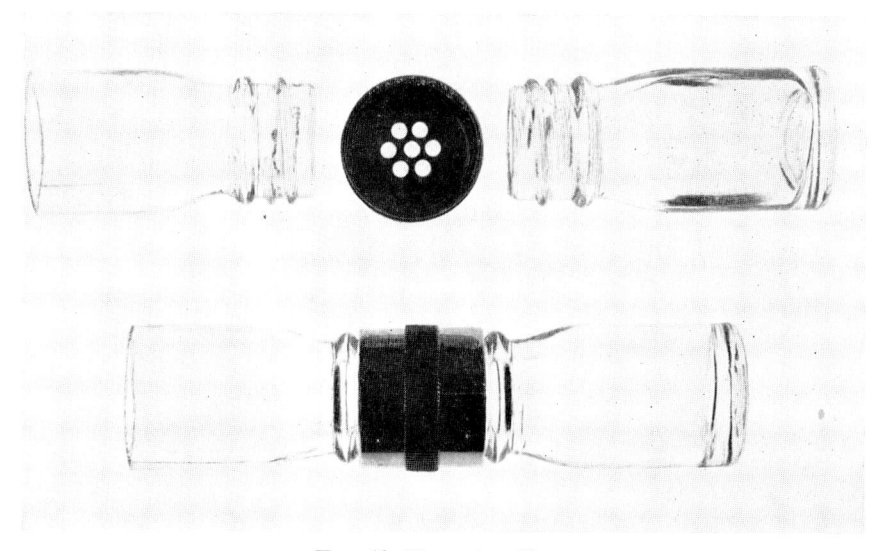

FIG. 62. Hemming filter.

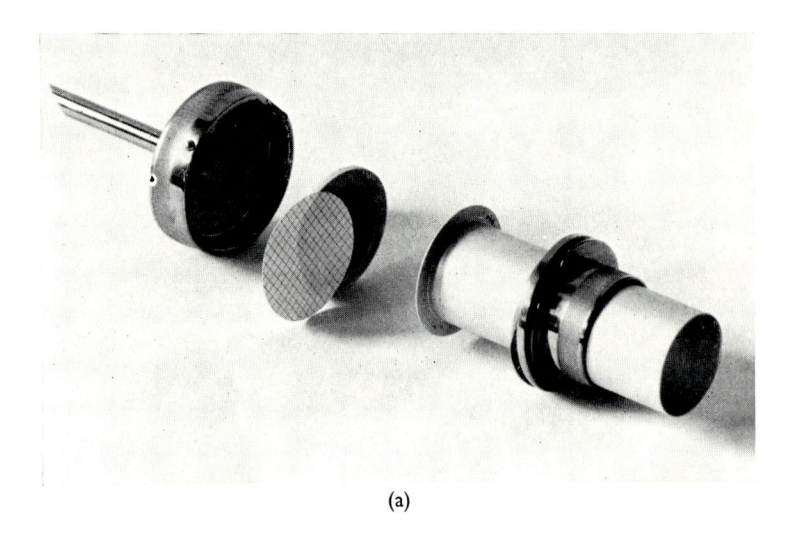

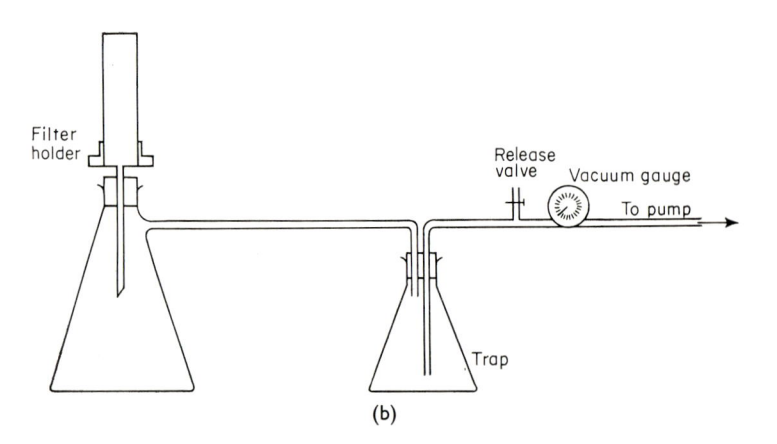

Fig. 63. (a), exploded view of filter holder, showing insertion of membrane filter; (b), assembly for pressure filtration.

For use, the apparatus is re-assembled with a membrane in position, and fitted to a filtration flask attached to a water-pump. A mercury manometer, or vacuum gauge, complete with release valve, is connected in series.

By using a negative pressure the membrane is primed with isotonic saline. Suction is removed while a residue of saline is in the filter cup; the membrane must not be allowed to dry. The pump is re-connected, and the fluid passed through, by using a negative pressure equal to 20 mm of mercury, and the last of the fluid is washed through with saline.

When filtration is complete, the membrane is removed, fixed in alcohol - ether solution for 20 min and stained with Papanicolaou's sequence. After the final xylene bath the membrane is cut and mounted between slide and coverslip by using DPX mountant.

(*iii*) *Notes.* Any clots present in the fluid must be removed before filtration.

The method has been used for various fluids, including sputa, after digestion with trypsin.

Fluids containing high concentration of cells (including erythrocytes) are unsuitable for this technique.

Tissue Culture

This term is used to describe methods that allow the *in vitro* propagation of somatic cells. Cultures are carried out to obtain preparations of a patient's cells in mitosis, so that their constituent chromosomes can be counted and examined. Counts are made by direct observation; assessment of morphology is accomplished by photographing suitable cells. Prints of the photographs are cut so that the individual chromosomes of a cell can be arranged in a systematic fashion. About thirty cells are analysed to prepare a report.

All the techniques follow a general pattern. Venous blood is the most convenient source of material for culture. Nucleated cells in the specimen are separated from the erythrocytes by using density-gradient techniques before addition to the culture medium. Committed cells, mainly neutrophils and eosinophils, quickly die, and the lymphocytes are the cells we see in mitosis. To induce these cells to divide a mitogenic agent, phytohaemagglutinin, is added. After the culture has been incubated for 3 days, an antimitotic agent, colchicine, is added. This substance has the property of arresting mitosis at metaphase: an accumulation of cells at this stage of division occurs, which is when the chromosomes may be most easily examined. The cells are harvested and spread on slides so that they do not overlap one another.

Macro-Method

In this method (Kernohan, 1963), collect about 10 ml of blood into a disposable polystyrene syringe previously rinsed out with heparin solution (see Fig. 64).

Holding the syringe upright, remove the needle and withdraw the piston to its extreme position, so that the level of the blood is almost 2 cm below the nozzle.

Heat the nozzle in a bunsen flame until the polystyrene begins to melt, and complete the seal by squeezing the nozzle with *cold* 5-in forceps.

When the seal is cool to touch, invert the syringe, remove the piston and stopper with a sterile rubber bung.

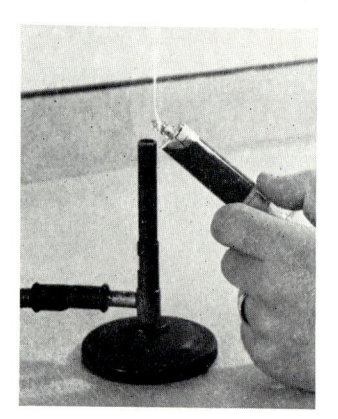

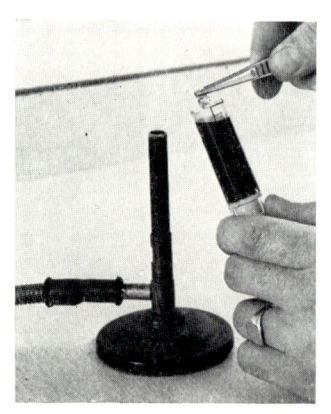

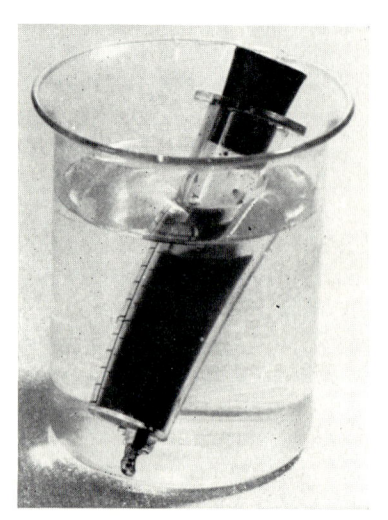

Fig. 64. Technique for sealing syringe, tissue culture macro-method.

Using a sterile pipette, add 0·2 ml of phytohaemagglutinin solution and mix.

Stand the syringe, nozzle downwards, in a beaker containing ice and water and leave for 30 min.

Aspirate the cell-rich plasma as three approximately equal volumes with disposable 10-ml syringes.

Add 4 volumes of TC199 solution to the volume of plasma in each syringe and incubate for 72 h at 37°C.

Three hours before harvesting, add accurately 0·1 ml of 0·004% colchicine for each 1 ml of culture, and mix.

Transfer the contents of the syringes into each of three clean centrifuge tubes and centrifuge at 300 rev/min for 5 min.

Remove the supernatant liquor and re-suspend the cells in a volume of distilled water at 37°C equal to the volume of supernatant liquor removed.

Incubate at 37°C for 20 min.

Centrifuge at 600 rev/min for 5 min and remove all but the last drop of supernatant liquor.

Re-suspend the cells, and carefully add 2 ml of fixative drop by drop gently mixing after each drop.

Stand the suspension at 4°C for 30 min.

Recover the cells by centrifugation and subsequently wash them with a further 2 ml of fresh fixative.

Centrifuge again and remove all but the last drop of supernatant liquor.

Re-suspend the deposit and, using a fine Pasteur pipette, add one drop of the suspension to a clean, wet, ice-cold slide tilted to drain off the surplus fluid. For method of cleaning slides see General Notes (*iv*), p. 130.

Dry by warming the slide in a flame: do not overheat.

Stain with lactic acid - orcein solution for 1 h, dehydrate and mount.

(a) *Solutions*

Heparin (Pularin; 5,000 units per ml, Evans Ltd).

Phytohaemagglutinin (Burroughs Wellcome Ltd). Dilute according to instructions.

TC199 solution, containing penicillin, streptomycin and sodium hydrogen carbonate (Glaxo Ltd).

Colchicine (Colcemid–Ciba Ltd). Colcemid contains 1 mg of colchicine per ml of solution. The working solution is prepared by taking 0·4 ml of Colcemid and diluting it to 1 ml with the TC199 solution (−0·004% colchicine).

Fixative. Mix 3 volumes of methanol with 1 volume of glacial acetic acid immediately before use.

(b) *Notes*

1. Instead of the 30-min immersion in ice - water, an alternative

method of obtaining the cell-rich plasma is to centrifuge the syringe, after sealing the nozzle, at 350 rev/min for 5 min.
2. The cell-rich plasma from leukaemic patients is diluted with AB serum to give cell counts of about 2,000/cu. mm before addition of the TC199 solution.
3. Poor preparations result if the pH of the culture, during incubation, falls below 6·8; this is seen as a change in colour of the medium from red to yellow.

Micro-Method

Collect peripheral blood into a 2-ml disposable syringe containing one drop of heparin. About 0·4 ml of blood can usually be obtained by the ear-prick method. Draw 1 ml of AB serum into the syringe and mix.

Add 0·01 ml of phytohaemagglutinin using a tuberculin syringe and mix (see Fig. 65).

Centrifuge the syringe (see General Notes (*iii*), p. 130) at 350 rev/min. for 5 min.

Dispense the cell-rich supernatant liquor (see Fig. 65 and General Notes (*iii*)) equally into three disposable 2-ml syringes, each containing 1 ml of TC199 solution.

Incubate for 72 h at 37°C.

Add 0·1 ml of colchicine solution at 37°C 4 h before harvesting.

Transfer the contents of each syringe to three clean, conical centrifuge tubes.

Centrifuge at 600 rev/min for 5 min.

Remove the supernatant liquor and add 0·5 ml of water at 37°C.

Incubate at 37°C for 10 min.

Cool in ice - water, add 2 ml of fixative, and mix.

Centrifuge, discard the supernatant liquor and re-suspend the deposit in 2 ml of fixative.

Centrifuge, and remove all but the last drop of fixative. Re-suspend the deposit.

Using a fine Pasteur pipette, add one drop of suspension to a cleaned, wet, ice-cold slide tilted to drain off surplus fluid.

Dry by passing quickly over a flame—do not overheat.

Stain with lactic acid - orcein solultion for 1 h, dehydrate and mount.

(a) *Solutions*

Heparin (Pularin; 5,000 united per ml, Evans Ltd).

Phytohaemagglutinin (Burroughs Wellcome Ltd). Dilute according to instructions.

AB serum.

TC199 solution (containing penicillin, streptomycin and sodium hydrogen carbonate (Glaxo Ltd).

Colchicine, 0·04% aqueous (BDH Ltd). Dilute 1 to 10 before use.
Fixative. Mix 3 volumes of methanol with 1 volume of glacial acetic acid immediately before use.

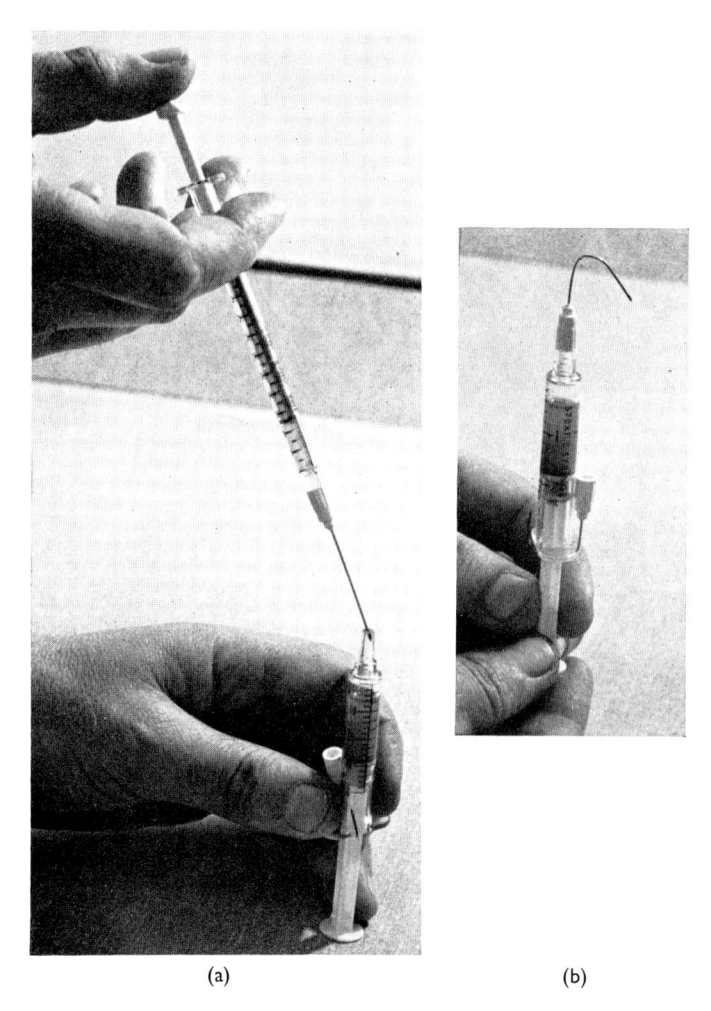

(a)   (b)

Fig. 65. Tissue culture micro-method: (a), addition of phytohaemagglutinin; (b), transfer of supernatant liquor (the lower, locking, needle is first withdrawn).

(b) *General Notes*

(*i*) Aseptic precautions, using sterile apparatus and solutions, are necessary until the 72-h incubation period is ended.

(*ii*) Disposable syringes eliminate the preparation of sterile siliconized glassware.

(*iii*) The plunger of the syringe is locked by passing a hot needle through the stem immediately above the barrel. After centrifuging, the pin can be withdrawn and the syringe contents aspirated by using a hypodermic needle bent on itself.

(*iv*) Special attention should be given to the preparation of the slides; grease should not be removed with detergents. A suggested method (Edwards, 1962) is as follows. Prepare a series of three Coplin jars containing: A, 1 volume of alcohol; 1 volume of N sodium hydroxide and 3 volumes of distilled water; B, N hydrochloric acid; C, distilled water. Leave slides in solution A for a period of time ranging from hours to days, in solution B for a period of time ranging from minutes to days; slides may be stored in solution C. The volume of liquid in the Coplin jars is adjusted so that the upper ends of the slides are clear of the solution and therefore remain dry and dirty. They can then be handled without contaminating the wet glass with grease from the fingers. Immediately before a slide is used, the surface film is renewed by squirting on distilled water.

(*v*) Before being prepared as permanent mounts, the specimens may be examined as wet films by using either 1% aqueous methylene blue, 1% aniline blue or the orcein solution as mountant.

Interpretation

The preparation of karyotype and idiogram is outlined under The Normal Cell, p. 28. Information accruing from these investigations has demonstrated chromosomal aberration in a variety of disorders. Significant findings are found, for example, in mongolism, when a particular chromosome is represented by three autosomes instead of the expected two. Most frequently the extra chromosome is an additional number 21, and the patient is said to exhibit trisomy 21. The most easily detected chromosome anomalies occur in the sex chromosomes, since the second, and any additional, X chromosomes can be identified directly in vesicular, interphase nuclei of many somatic cells (see p. 131).

The sex karyotype of a normal chromosomal male is XY, and that of a female XX. Deviations that can occur are the absence of one or other of the sex chromosomes, when the missing chromosome is indicated with a zero, i.e., XO or YO, and the presence of additional sex chromosomes, e.g., in females, triplo-X (XXX) or tetra-X (XXXX), and in males, diplo-X, Y (XXY) triplo-X, Y (XXXY).

For an account of chromosomal abnormalities, see Moore and Hey (1963).

Chromosomal (Nuclear) Sex

Cytological assistance in the diagnosis of errors in sex development (Klinefelter's and Turner's syndromes) can be offered without recourse to tissue culture and karyotyping of the patient's somatic cells. It has been shown that the presence of two or more X chromosomes can be established by direct microscopy of the interphase nucleus. As has been explained (see Tissue Culture, p. 125), the sex complement of genetically male cells would be one X chromosome and one Y chromosome: a genetically normal female cell would carry two X chromosomes.

The first X chromosome in the interphase nucleus uncoils, as do the autosomes, and becomes lost in the chromatin reticulum. Other X chromosomes that may be present remain condensed and can be readily recognised in the interphase nucleus as darkly staining ovoid masses (Barr bodies), about 0.7 to 1.2 μ, contiguous with the nuclear rim (see Fig. 66). A genetically normal male will have cells containing no such bodies and a normal female will have cells containing one. Triplo-X and tetra-X females will exhibit two and three X chromatin masses, respectively, in their nuclei.

Squamous and columnar epithelial cells exhibit these bodies, as do other somatic cells. A commonly chosen site for collecting cells for this investigation is the buccal mucosa. The mouth is first thoroughly rinsed with water to remove debris and bacteria. Material is then collected by scraping the mucosa with a wooden tongue depressor, which is then lightly drawn over a clean, glass slide. After fixation with alcohol - ether, the smear is stained with cresyl fast violet (see Staining, p. 18).

For reasons not yet understood, not all the cell nuclei of a normal female possess the Barr body. A differential count of the cells may show only 30% of cells to contain the structure. A smaller chromatin mass, the male chromocentrum, may be seen in a few nuclei (about 5%) of squamous cells from normal males.

Neutrophils Accessory Nuclear Lobule (Drum Stick)

Sex differences can also be shown in mature neutrophils. When these cells are examined in stained blood films from females, a proportion (0.6 to 8.8%) are seen to possess a characteristic, "drum stick" shaped appendage (see Fig. 66a). The structure is not seen in neutrophils of normal males.

A helpful account of sex-chromatin determinations is given by Bamford *et al.* (1963).

Detection of Foetal Squamous Cells

Occasionally, during the latter part of a pregnancy, the amniotic membrane may rupture several days before delivery with a consequent escape of the contained fluid into the vagina. The presence of foetal material in the vaginal pool may be used as an indicator to detect this (Kittrich, 1963). The evidence comprises squamous cells shed from the foetus and also a foetal sebaceous secretion (*vernix caseosa*).

Smears from the vaginal pool, received unfixed, are stained by adding one or two drops of 1% aqueous nile blue sulphate. The preparation is mounted with a large coverslip, which covers the smear, and sealed with Glyceel (G. T. Gurr Ltd). Maternal squamous cells are stained blue, whereas the foetal cells and the fat globules stain orange. A rapid examination of the slide with the ×10 objective will reveal the foetal material if present.

Control smears of known positive material can be stored for several months. The slides are placed, unfixed, in airtight containers that are kept at 5°C.

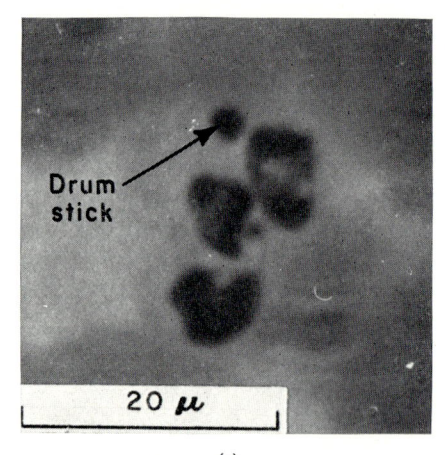

(a)
Neutrophil showing drum stick.

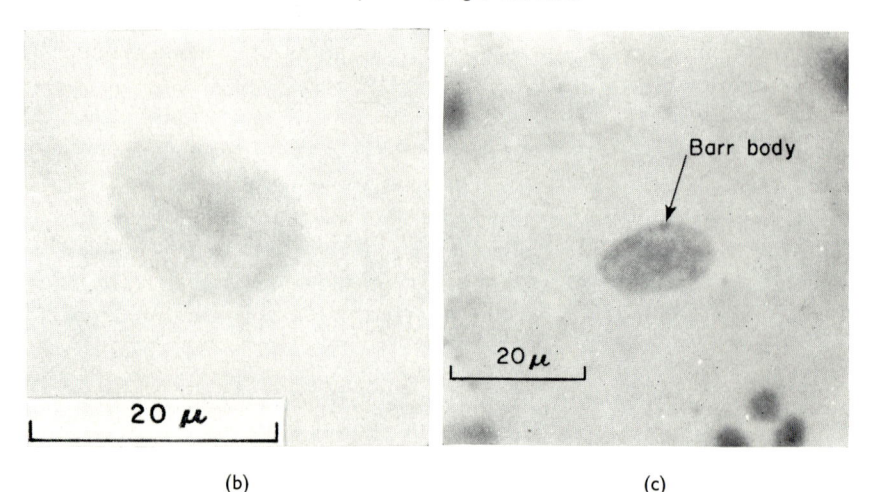

(b) Squamous cell nucleus without Barr body.   (c) Squamous cell nucleus with Barr body.

FIG. 66.

Appendix

Formulae

Haematoxylins

(a) *Harris's*

Haematoxylin	1 g.
Alcohol, 74 O.P.	10 ml.
Potassium alum	20 g.
Distilled water	200ml.
Mercuric oxide	0·5 g.
Acetic acid, glacial	8 ml.

Dissolve the haematoxylin in the alcohol, the alum in the water and mix together. Heat until just boiling, remove from heat and add the mercuric oxide. This causes the mixture to change to a purple colour. Cool immediately in a cold-water bath. When cool, add the acetic acid. The solution is now ready for use.

(b) *Ehrlich's*

Haematoxylin	2 g.
Alcohol—74 O.P.	100 ml.
Distilled water	100 ml.
Glycerine	100 ml.
Acetic acid, glacial	10 ml.
Potassium alum	Saturate.
Sodium iodate	0·4 g.

Prepare the alcoholic haematoxylin solution and add it to the rest of the mixture. Warm to 60°C and add the sodium iodate. The solution is now ready for use.

(c) *Weigert's*

Solution A:

1% haematoxylin in alcohol 74 O.P.

Solution B:

Ferric chloride	4 ml.
Distilled water	95 ml.
Hydrochloric acid, conc.	1 ml.

Immediately before use mix equal volumes of the two solutions.

Scott's Tap Water Substitute

Sodium bicarbonate	3·5 g.
Magnesium sulphate	20 g.
Tap water	1 litre

Do not add a preservative. If the solution becomes contaminated with moulds, discard it.

Acid Alcohol

Alcohol, 74 O.P.	300 ml.
Distilled water	100 ml.
Hydrochloric acid, conc.	2 ml.

Eosin

Eosin BWS	5 g.
Distilled water	100 ml.

Dilute 1 in 5 with water for use.

Papanicolaou's Stain

These stains are best purchased as commercially prepared solutions.

(a) *O.G.6*

Orange G crystals	10 g.
Distilled water	100 ml.
Alcohol, 74 O.P.	1 litre.
Phosphotungstic acid	0·15 g.

(b) *EA50 (EA36)*

Light green SF solution, 0·5% in 95% alcohol	45 ml.
Bismark brown solution, 0·5% in 95% alcohol	10 ml.
Eosin yellow solution, 0·5% in 95% alcohol	45 ml.
Phosphotungstic acid	0·2 g.
Lithium carbonate sat. aq. solution	0·05 ml.

(c) *EA65*

Light green SF solution, 0·25% in 95% alcohol	45 ml.
Bismark brown solution, 0·5% in 95% alcohol	10 ml.
Eosin yellow solution, 0·5% in 95% alcohol	45 ml.
Phosphotungstic acid	0·2 g.
Lithium carbonate, sat. aq. solution	0·05 ml.

Schorr's Stain

Alcohol 74 O.P.	100 ml.
Biebrich scarlet (WS)	0·5 g.

Orange G	0·25 g.
Fast green FCF	0·075 g.
Phosphotungstic acid	0·5 g.
Phosphomolybdic acid	0·5 g.
Acetic acid, glacial	1·0 ml.

Guard's Stain

Biebrich scarlet (WS)	1·0 g.
Phosphotungstic acid	0·3 g.
Acetic acid, glacial	5·0 ml.
Alcohol 50%	100 ml.
Fast Green FCF	0·5 g.
Phosphomolybdic acid	0·3 g.
Phosphotungstic acid	0·3 g.
Acetic acid, glacial	5·0 ml.
Alcohol 50%	100 ml.

Unna–Pappenheim (Jordan and Baker)

0·5% aqueous methyl green	13 ml.
0·5% aqueous pyronin Y	37 ml.
Acetate buffer, pH 4·8	50 ml.

The stain is best prepared afresh every 4 weeks from the component stock solutions, which keep for about 6 months. The stock methyl green is washed free from methyl violet after preparation by successive extractions with chloroform until a colourless wash is obtained.

Acetate buffer, pH 4·8

Acetic acid, 0·1 N	80 ml.
Sodium acetate, 0·1 N	120 ml.

Prepare 0·1 N acetic acid by diluting 6 ml of glacial acetic acid to 1 litre with distilled water.

Prepare 0·1 N sodium acetate by dissolving 13·6 g of the hydrated salt in 1 litre of distilled water.

Ribonuclease Solution

Crystalline ribonuclease	10 mg.
Distilled water	10 ml.

The solution is stored at 0°C when not in use.

Lactic Acid - Orcein

Synthetic orcein	1 g.
Acetic acid, glacial	45 ml.

The mixture is boiled to dissolve the orcein: on cooling, 40 ml of 70% lactic acid is added and the solution filtered.

Aniline Blue

Aniline blue	1 g.
Acetic acid, 1%	100 ml.
Filter.	

Cresyl Fast Violet

Cresyl fast violet	1 g.
Acetic acid, 0·01 N	100 ml.
Filter.	

Nile Blue Sulphate

Nile blue	0·1 g.
Distilled water	100 ml.
Filter.	

Romanowsky Stains

Suitable stock stains are available from G. T. Gurr Ltd, London.

Leishman

Stock powder	0·15 g.
Methanol, A.R.	100 ml.

Wash the weighed powder into a bottle with the methanol. Stopper the bottle and allow the solution to stand for 4 days at room temperature with occasional shaking. Filter the solution.

For use the stain is diluted with an equal volume of buffer solution, pH 6·8. (For buffer solution, pH 6·8, see below.)

Jenner's Stain

Stock powder	0·5 g.
Methanol, A.R.	100 ml.

Prepare as for Leishman. For use, dilute one volume with two volumes of buffer solution pH 6·8.

Geimsa's Stain

Stock powder	2·8 g.
Methanol, A.R.	375 ml.
Glycerol	125 ml.

Dissolve the powder as for Leishman. For use dilute one volume with eight volumes buffer solution pH 6·8.

APPENDIX

Buffer solution, pH 6·8

Sodium phosphate, M/15	49·6 ml.
Potassium phosphate, M/15	50·4 ml.

Prepare M/15 sodium phosphate by dissolving 9·465 g of disodium hydrogen phosphate (Na_2PHO_4) in 1 litre of distilled water, and prepare M/15 potassium phosphate by dissolving 9·070 g of potassium dihydrogen phosphate (KH_2PO_4) in 1 litre of water.

Phosphate buffer in tablet form is obtainable from G. T. Gurr Ltd, London.

Feulgen's Stain

(a) *Schiff's reagent*

Basic fuchsin	1 g.
Distilled water	400 ml.
Thionyl chloride	1 ml.

Dissolve the stain by boiling it in the distilled water. Cool and filter the solution. Add the thionyl chloride carefully drop by drop. Stand the solution overnight in the dark. Clear the solution with activated charcoal, filter and store it in the refrigerator.

(b) *Reducing agent*

Potassium metabisulphite	1 g.
Distilled water	200 ml.
Hydrochloric acid, N	10 ml.

(c) *Tartrazine*

Tartrazine in Cellosolve	Saturated solution.

Albumen Adhesives

(a) *Egg albumen*

Glycerine	1 volume.
Distilled water	1 volume.
Egg albumen	1 volume.

Mix the constituents and filter the mixture through a fast paper (Whatman No. 4 or 41).

(b) *Bovine albumen*

Albumen, 30%	1 ml.
Distilled water	9 ml.

Mix and filter.

DPX Mountant (Kirkpatrick and Lendrum)

Distrene 80	10 g.
Dibutyl phthalate	5 ml.
Xylene	35 ml.

Mix and store in a stoppered bottle.

Industrial Spirit, 74 O.P.

In order to demonstrate awareness of the limitations suffered by this reagent in respect of its substitution for ethanol ("absolute alcohol"), candidates are sometimes questioned about its constitution. A typical analysis, conforming to British Standard Specification (BS 3591: 1963) would be:

Total alcohols, 99·2% w/w
Water, 0·8% w/w
  and containing
Methanol, 4·8% w/w
Aldehydes and ketones (as acetaldehyde), 0·10% w/w
Acidity (as sulphuric acid), 0·002% w/w
Residue on evaporation, 0·002% w/w
Miscibility with water complete.

The reagent is obtained by blending 95 volumes synthesized ethanol with 5 volumes wood naphtha. Contaminants originate in the added wood naphtha, and of these the most important are methanol and the reducing substances (aldehydes and ketones, largely present as acetone).

Note on Centrifuges

Consideration of Stoke's Law:

$$S_r = \frac{2a^2(\rho_p - \rho_m)g}{9\eta}$$

where S_r is the rate of settling, a is the radius of the particle, ρ_p is the density of the particle, ρ_m is the density of the medium, g is the acceleration due to gravity (981 cm sec^{-2}) and η is viscosity of the medium, shows that the rate of settling is increased by:

(a) increase in size of the particle

(b) increase in the difference between the density of the particle and the density of the medium

(c) decrease in the viscosity of the medium

(d) increase in the acceleration due to gravity.

In biological investigations, it is rarely possible to vary the factors (a) to (c). We can, however, substitute a greater force for g by using the centrifuge.

In the steady state, the centrifugal force generated by a centrifuge is a function of not only the speed of rotation of the head but also of the radius of the arms. The conditions for centrifuging in many medical laboratory techniques are frequently given by defining only the time and the speed

APPENDIX

(usually in rev/min), mainly because centrifuges of "standard" type are commonly used and the centrifugal force is not absolutely critical. A "standard" centrifuge has arms of radius 15 cm and no account is taken of the differences caused by air turbulence in varying-shaped bowls. This convention has been followed in this book when techniques involving centrifuging are described. The relative centrifugal force (RCF) in gravitational units can be obtained by applying the formula:

$$RCF = 1\cdot118 \times 10^{-5} \times 15 \times N^2$$

where N is the speed in rev/min.

Derivation of Formula

The centrifugal force on a particle moving in a horizontal circle is given by:

$$\text{Force} = \text{mass} \times \text{acceleration}$$

i.e. $F_c = m \dfrac{\omega^2}{r}$

where F_c is the centrifugal force, m is the effective mass of the particle, ω is the angular velocity and r is the radius of the circle.

The angular velocity, ω, for a particle rotating at n rev/sec is given by:

$$\omega = 2\pi n$$

Therefore:

$$F_c = \frac{m(2\pi rn)^2}{r} = 4m\pi^2 rn^2$$

The gravitational force, F_g, on a particle of mass m is given by:

$$F_g = mg$$

Thus:

$$\frac{F_c}{F_g} = \frac{4m\pi^2 rn^2}{mg} = \frac{4\pi^2 rn^2}{g}$$

Expressing rev/sec (n) in rev/min (N) and substituting for π, we obtain:

$$\frac{F_c}{F_g} = \frac{4 \times 3\cdot1416^2 \times r \times N^2}{981 \times 60 \times 60}$$
$$= 1\cdot118 \times 10^{-5} \times r \times N^2.$$

Example

A centrifuge, radius 10 cm, speed 4000 rev/min has:

$$RCF = 1\cdot118 \times 10^{-5} \times 10 \times 4000^2$$
$$= 1788\cdot8 \sim 1800g$$

The time taken for a particle, having a settling time of 120 min by gravity alone, to settle in such a centrifuge is given by:

$$\frac{120}{F_c} \text{ min} = 4 \text{ sec.}$$

Note on Population Screening for Cancer of the Cervix

The Ministry of Health has recognised the need for a screening programme for "apparently well" women, outlining its policy in publication H.M. (66)76.

It offers guidance for a planned approach suggesting an initial service for women age 35 years and over at 5 year intervals subsequently expanding to cover all women at risk.

Centralization of cytology laboratory facilities is recommended, the laboratories to be staffed by pathologists and medical laboratory technicians trained in cytology. The capacity of these laboratories is to be assessed at 7000 smears/worker annually.

A planning committee to co-ordinate development is suggested comprising representatives from local Hospital Authorities, Executive Councils, Local Medical Committees and Local Health Authorities, but priority is acknowledged for hospital gynaecological and obstetrical patients. Smears should be collected by doctors and nurses trained in the techniques, and hospital gynaecological departments are invited to offer instruction to these workers.

Free comprehensive kits are available from the Ministry's Supply Division S(B)2B London by requisition on a Cytology Order Form. The kits contain Ayre's spatulae, slides, containers, fixative (alcohol+ water-soluble wax) and request forms. General practitioners have access to the service by completing a cervical cytology certificate, Form E.C.74, to cover remuneration and request form, H.M.R. 101/5, to accompany the specimen.

Provision for the necessary clerical staff and storage space should be made by Hospital and Local Health Authorities, and publicity for the service controlled to regulate the demand.

Bibliography

Abramson, W., Dzenis, V., and Hicks, S. (1964). *Acta cytol.* **8**, 306.
Anderson, W. A. D. (1961). "Pathology." Henry Kimpton, London.
Bamford, S. B., Cassin, C. M., and Mitchell, B. S. (1963). *Acta cytol.* **7**, 151.
Bertalanffy, B. D., and Bull, C.A. (1960). *Am. Cancer Soc.* **10**, 118.
Boddington, M. M., Spriggs, A. I., and Wolfendale, M. R. (1965). *Br. med. J.* (i), 154.
Chapman, G., and Barker, W. B. (1964). "Zoology for Intermediate Students." Longmans, London.
Davies, H. J. (1962). *Am. J. Obstet. & Gynec.* **84**, 1017.
Edwards, J. H. (1962). *Cytogenetics*, **1**, 90.
Graham, R. M., and Graham, J. B. (1953). Cancer, *Philad.* **6**, 215.
Graham, R. M., and Graham, J. B. (1955). *Cancer, Philad.* **8**, 59.
Graham, R. M. (1963). "The Cytologic Diagnosis of Cancer." W. B. Saunders & Co., London.
Kasdon, S. C., and Bamford, S. B. (1962). "Atlas of *in situ* Cytology." Churchill, London.
Kjellgren, O. (1964). *Acta cytol.* **8**, 216.
Kernohan, W. (1963). Personal communication.
Kittrich, M. (1963). *Geburtsh. Franenheilk.* **23**, 156.
Kohn, J., and Earl, J. M. B. (1965). *J. clin. Pathol.* **18**, 479.
Koss, L. G., and Durfee, G. R. (1961). "Diagnostic Cytology and Its Histologic Bases." Pitman Medical Publishing Co., London.
Long, N. L. (1963). *Acta cytol.* **7**, 85.
Maximow, A. A., and Bloom, W. (1948). "A Textbook of Histology." W. B. Saunders & Co., London.
McInroy, R. A. (1966). *Br. med. J.* (i), 1472.
Meissals, A. (1963). *Acta cytol.* **7**, 239.
de Mello, R. P. (1963). *Acta cytol.* **7**, 62.
Moore, K. L., and Hay, J. C. (1963). *Can. med. Ass. J.* **88**, 1022, 1071.
Naylor, B. (1964). *Acta cytol.* **8**, 141.
Papanicoloaou, G. N. (1954). "Atlas of Exfoliative Cytology." Harvard University Press, Cambridge Mass.
Philps, F. R. (1964). "A Short Manual of Respiratory Cytology." Pitman, London.
Saccomano, G., and Saunders, R. P. (1963). *Acta cytol.* **7**, 305.
Schade, R. O. K. (1959). *Acta cytol.* **3**, 7.
Seal, S. H. (1956). *Cancer, N.Y.* **9**, 866.
Seal, S. H. (1959). *Cancer, N.Y.* **12**, 590.
Sharpe, H. S. (1963). *J. med. Lab. Techn.* **20**, 245.
Shaw, S. L. (1965). *J. med. Lab. Techn.* **22**, 139.
Smolka, H., and Soost, H. J. (1965). "An Outline and Atlas of Gynaecological Cytodiagnosis." Edward Arnold Ltd, London.
Umiker, W. (1964). *Acta cytol.* **8**, 186.
Wakonig-Varrtaja, R., and Hughes, D. T. (1965). *Lancet*, (2), 756.

Subject Index

A

Acetate buffer solution, formula and use, 135
Acid alcohol, 12
  formulae, 16, 134
Acridine Orange, 7, 96, 115
Adam's endometrial brush, 70
Albumen adhesives, 137
Alcohol
  in fixatives, 2
  in sputum collection, 86
Alcohol - acetic acid fixative, 3
Alcohol - ether fixative, 3
Alveolar cells, 39
Anaphase, 33
Aniline blue chromosome stain, 130
  formula, 136
Anisonucleosis, 51
Artist's Picture Varnish, 22
Ascitic fluid, 93
Atrophic epithelium, 74
Atypical epithelial cells, 44
Autofluorescence, 113
Automatic staining, 13
Autosome, 27
Ayre spatula, 69

B

Barr bodies, 131
Basal cells, 41
Basement membrane, 35
Basophils, 62
Biebrich Scarlet - Fast Green (Guard's) stain
  method, 17
  solutions, 135
Bird's eye cell, 58
Blocking technique, rapid wax embedding, 117
Blood
  atypical cells in, 96
  in gynaecological smears, 71
  preparation of smears, 96
Blueing smears, 12
  Scott's tap water substitute, 134
Bovine albumen adhesive, 137

Breast secretions, 103
Buffer solution
  Acridine Orange stain, 115
  phosphate pH 6·8, 127
  Unna-Pappenheim stain, 135

C

Carnoy's fixative, 4
Cell
  epithelial, columnar, 42
            squamous, 39
            transitional, 41
  general, 25
  reproduction, 26 *et seq.*
  shape, 39
  structure, 25
Cellophane
  use in cytology, 1, 13, 22
  use in histology, 120
Centrifuge, 138
Centrosome, 25
Cerebrospinal fluid (CSF), 104
Cervical smear, normal, 73
Cervicitis, 78
Cervix, epithelium of, 68
Chiasmata, 34
Chromatid, 27
Chromatin, 25
  Atypical changes, 50 *et seq.*
Chromocentrum, male, 131
Chromomere, 28
Chromosomal sex, 131
Chromosome, 27 *et seq.*
  classification, count, description, idiogram, number, sex, and also structure, 131
Cilia, 43
Cleaning smears, 13
Clearing tissue, rapid method, 116
Coating solution for "dry" fixation, 6
Coding information, 105
Coiling (of chromosome), 32
Colchicine, in tissue culture, 129
Cold agglutinins, in serous fluid, 93
Collection of material
  blood, 96
  gastric juice, 99

Collection of material—*continued*
  gynaecological, 68 *et seq.*
  nipple exudate, 103
  prostatic secretion, 104
  serous fluids, 93
  sputum, 86
  urine, 102
Columnar cells, 35, 42, 43
Concentration techniques, for cells, 121
Cornification, of squamous cell, 41
  Index, 76
Coverslips, 20
  substitute for, 22
Crenation of nucleus, 52
Cresyl fast violet
  method, 18
  formula, 136
CSF (Cerebrospinal Fluid), 104
Cubical epithelial cells, 42
Cuboidal epithelial cells, 35, 41
Cyanophilia, 41
Cytofixative, Ortho, 4
Cytological fixatives, 3
Cytolysis, 53
Cyto-microanatomical fixatives, 118
Cytopipette, Davis, 68
Cytoplasmic inclusions, 40, 53, 63

D

Davis cytopipette, 68
Data retrieval, 107
Dehydrating smears, 10
  tissue (rapid method), 116
Denver system, of chromosome classes, 2
Designated cells, 56
Differentiation, in staining, 12
Differentiating smears, 12
Diploid number, 26
Desoxyribonucleic acid (DNA), 27
  Staining methods, 19, 20
Documentation, 105
DPX Mountant, formula, 137
"Drumstick"(accessory nuclear lobule), 131
"Dry" fixation, 4
  coating solution, 6
Drying effect, 6
Dyskaryosis, 45
  in moniliasis, 81
  in trichomoniasis, 82
Dysplasia, 45

E

EA 50, Formula, 134
EA 65, Formula, 134
Ectocervix, 68
Ectoderm, 34
Effusions pleural, 93
Ehrlich's haematoxylin
  method, 16
  formula, 133
Endoderm, 35
Endometrial brush, 70
Endometrium, 68
Endothelial cell, 35
Endothelium, 35
Envelope Effect, in squamous cells, 77
Eosin, formula, 134
Eosinophil, 62
  description of, 66
Eosinophilic index, 76
Epithelial cells
  classification, 35
  normal, 38
  squamous, 39
Epithelial pearl, 56
Epithelial tissue, 34, 35
Epithelium
  pseudostratified, 37
  simple, 35
  stratified, 35
  transitional, 37
Erosion of epithelium, 46
Erythrocyte, 62
  description, 67
Examination of smears, 22

F

FCF fixative, 4
Feulgen stain
  method, 19
  solutions for, 137
Fibre cell, 56
Fibrinogen, malignant cells in blood, 96
Fibroblast, 62
  description, 65
Filters
  fluorescence, 113
  Hemming, 123
  pressure, 124
Filtration, method for preparation cell concentrates, 122

Fixation
  of cells, 2
  of tissues, 116
Fixatives
  alcohol/acetic, 3
  alcohol/ether (Papanicolaou's), 3
  alcohols in, 2
  Carnoy's, 4
  commercial, Cytofixative, 4
  Spraycyte, 4
  "dry", 6
  formol/alcohol, 3
  for tissue processing, 118
  methanol, 4
  recovery, 4
  Schaudin's, 3
Flotation technique, 122
Fluorescence microscopy, 112 et seq.
Fluorochrome, 113
Foetal squames, detection of, 132
Folding
  of cytoplasm, 407
  of nucleus, 52
Foreign body giant cell, 63
Formalin/alcohol, 3
Formulae
  adhesives, 137
  buffers, 137
  industrial spirit, 138
  mountant, 137
  stains, 133

G

Gametogenesis, 26
Gastric juice, 97
  collection, 99
  cytology, 97
  preparation of smears, 100
Geimsa stain, 136
Gene, 28
General cell structure, 25
Genital tract, epithelium of, 68
Giant cell, 63
Giant histiocyte, 63
Golgi zone, 26
Grading of cells, 58
Gram stain, 17
Granulocyte, 62
Guard's Stain, 17
  formula, 135
Gynaecological cytology, 68 et seq.

H

Haematoxylin
  Ehrlich's, 16
  formulae, 133
  Harris's, 15
  Weigert's, 16
Haematoxylin and Papanicolaou sequence, 15
Haematoxylin - eosin sequence, 16
Haploid number, 26
Harris's haematoxylin, 15
  Formula, 133
Helical coiling, of chromosomes, 32
Hemming filter, 123
Heparin
  in tissue culture, 126, 128
  in blood smears, 96
Histiocytes
  description of, 62
  in sputum, 90
Hof nuclear, 62
Hydrating smears, 10
Hyperchromia, 50
  in cuboidal cells, 42
Hyper-oestrinism, appearance of smear in, 74
Hypochromia, nuclear, 50
Hypo-oestrinism, appearance of smear in, 74

I

Idiogram, chromosome, 29
Inclusions
  columnar cell, 00
  cytoplasmic, 26, 53
  histiocyte, 63
  squamous cell, 40
Indices
  cornification index, 76
  eosinophilic index, 76
  maturation index, 76
  reproducibility of, 77
Industrial spirit, 2
  constituents of, 138
Infection seen in gynaecological smears
  monilia, 79
  trichomonas, 81
Inflammation, 46
Inflammatory smear, gynaecological, 77
In Situ carcinoma, 45

146 PRACTICAL CYTOLOGY

Interchromomere, 28
Intermediate squamous cells, 41
Interphase, 33
Isobutanol, 2
Isopropanol, 2
    fixative for sputum, 86

J

Jenner–Giemsa stain, 19

K

Karyolysis, 52
Karyorrhexis, 52
Karyotype, 29
Keratin, 37
Keratinization, 37, 41

L

Labelling slides, 21
Lactic acid orcein
    solutions for, 135
    stain, 127
Lactobacillus, 73
Laking, of chromatin, 50
Leishman stain, 18
Leptothrix in trichomoniasis, 83
Lymphocyte, 62
    description of, 66

M

Macrophage, 62
Macroscopic examination of smears, 70
Mailing smears, 6
Malignancy, criteria for, 47
Malignant cells
    general criteria, 47
    in blood, 96
    in gastric aspirates, 99
    in gynaecological smears, 83
    in sputa, 91
    in urine, 102
    Papanicolaou grades, 58
Marking cells, 24
Maturation index, 76
Meiosis, 26, 34
Membrane
    cell, 25

cellulose, 123
nuclear, 25
Menstruation, changes in, 75 *et seq.*
Mesenchymal epithelium, 35
Mesoderm, 34
Mesothelial cells, in serous fluid, 94
Mesothelium, 35
Metachromal rhythm, 37
Metaphase, 32
Metaplasia, 44
Methanol fixation, 4
Methylene blue, chromosome stain, 130
Monilia infection, 79
Mounting smears, 20
Microscope
    positioning, 23
    scanning technique, 23
    suggested specification, 22
Mitochondria, 25
Mitosis, 26, 31
Mitotic cycle, 29
Mitotic figure, 56
Moniliasis, in gynaecological smears, 79
Morphological changes, classification of epithelial cell, 48
    cytoplasm, 53
    nuclei, 48
    table, 49
Moulding, in oatcell carcinoma, 92
Multinucleation, 51
    in cuboidal cells, 42

N

Naked nuclei, 46. 51
Navicular cells, 78
Neoplasia definition, 45
Neutrophil, 62
    accessory lobule, 131
    adherent, 53
    description of, 66
    inclusions, 53
Nile Blue stain
    in detection foetal squames, 131
    formula, 136
Non-epithelial cells in smears, 61
Normal smear
    cerebrospinal fluid, 104
    cervix, 73
    gastric aspirate, 97
    serous fluid, 94

sputum, 90
urine, 188
vaginal pool, 72
Nuclear/cytoplasmic ratio, 47
Nuclear changes, 48
Nuclear halo, 53
Nuclear membrane, 50
Nuclear sap, 26
Nuclear sex, 131
Nucleolus, 26, 51
Nucleotide, 27
Nucleus, general cell, 25

O

Oatcell carcinoma, 91
Objective, choice of microscope, 22
Oestradiol curve, 77
Oestrogens
  Effect on Vaginal Epithelium, 73
  Grading, 75
Oestrone grading, 75
OG 6 formula, 134
Orcein, chromosome stain, formula, 135

P

Papanicolaou
  fixative (alcohol/ether), 3
  grades, 58
  stain, 15
Papillary process, 43
Parabasal squamous cells, 41
Phagocytosis, histiocytes, 62
Phragmoplast, 33
Phytohaemagglutinin, in tissue culture, 125
Plaquarding, 78
Plasma cell, 62
  description of, 65
Pleomorphism, 37
Pleural fluid, 93
Polyps, 46
Postage of smears, 6
Pregnancy
  gynaecological smear in, 78
  macroscopic appearance of slide, 71
Primary fluorescence, 113
Progesterone effect, 77
Prophase, 32

Prostatic secretion, 104
Pseudo-giant cell, 63
Pseudo-stratified columnar epithelium, 37
Pyknosis, 33, 40, 52, 76

R

Radiation changes in cells, 84
Radiation response, 85
Rapid wax embedding, 116 et seq.
Reference cells
  collection, 109
  marking, 24
Rehydration of smears, 10
Re-staining smears, 13
Request forms, 105
Reticulo/endothelial cells, 62
Ribonuclease solution
  solution, 135
  use, 7, 20
Romanowsky Stains, formulae, 136
Rouleaux Effect, in T.V. Infection, 70
RR Test, 85

S

Satellite chromosome, 28
Scanning technique, 23
Schaudin's fixative, 3
Schiff's reagent, 137
Scott's tapwater substitute, formula, 134
Secondary fluorescence, 113
Senile vaginitis, 74
Sensitivity response, 85
Serial sections, storage of, 119
Serous fluids, 93 et seq.
Shorr's stain, 16
  formula, 134
Signet ring cell
  description and types, 56
  in gastric juices, 97
Silicone fluids, 122
Simple epithelium, 35
Skeining, 82
Slide carriers, 9
  postal, 6
Slides
  labelling, 21
  marking, 1

Smears
  fixation, 2
  mailing, 6
  preparation, 1
Spindle, 32
Spraycyte fixative, Clay-Adams, 4
Sputum
  collection, 86
  cytology of, 87
  malignant cells in, 91
  preparation of smears, 87
Squamo-columnar junction, 68
Squamous cells
  basal, 41
  intermediate, 41
  parabasal, 41
  superficial, 35, 39
SR test, 85
Staining
  automatic, 13
  differentiation, 7
  haematoxylin, 7
  romanowsky, 7
  technique, 9
Staining sequences, 15 *et seq.*
Staining technique, 9
Stains
  Acridine Orange, 115
  Ehrlich, 16
  Feulgen, 19
  fluorescent, 115
  Gram, 17
  haematoxylin, 16
  Harris, 15
  Jenner–Giemsa, 19
  Leishman, 18
  Romanowsky, 17
  Unna–Papenheim, 20
  Weigert, 16
Standards, cell size, 24
Storage of slides, 109
Stratified epithelium, 35
Stroma, 68
Superficial squamous cells, 40
Surveillance chart, 108
" Susa " fixative formula, 119

T

Tadpole cell, 56
Tartrazine stain formula, 137
TC 199, formula

  use in tissue culture, 127
Telophase, 33
Thionyl chloride, 137
Tissue culture
  macro-method, 126
  micro-method, 128
Toluidine Blue stain, for urine deposit, 103
Transitional cells in urine, 108
Transitional epithelium, 37
Trichomonas vaginalis, 81
Triton WR 1339 in cell flotation, 122

U

Ulcer, 46
Unna–Pappenheim stain
  formula, 135
  method, 20
Urine
  collection, 102
  cytology, 100
  preparation of smears, 102

V

Vacuole, 43
Vacuolation, cytoplasmic, 54
Vaginal, normal smear, 71
Varnish, substitute for coverslip, 22
Vernix caseosa, 132

W

Wax impregnation, rapid method, 117
Washing smears, 10
Weigert iron haematoxylin, 16
  formula, 133

X

X chromosome, 130, 131

Y

Y chromosome, 130, 131

Z

Zenker's fixative, 118
Zygote, 26